DEATH

— *A Reader* —

DEATH

—— *A Reader* ——

Mary Ann G. Cutter

University of Notre Dame Press

Notre Dame, Indiana

University of Notre Dame Press
Notre Dame, Indiana 46556
undpress.nd.edu
All Rights Reserved

Copyright © 2019 by University of Notre Dame Press

Published in the United States of America

Library of Congress Cataloging-in-Publication Data
Names: Cutter, Mary Ann Gardell, author.
Title: Death : a reader / Mary Ann G. Cutter.
Description: Notre Dame : University of Notre Dame Press, [2019] | Includes
  bibliographical references.
Identifiers: LCCN 2019011956| ISBN 9780268100520 (hardback : alk. paper) |
  ISBN 0268100527 (hardback : alk. paper) | ISBN 9780268100537 (pbk. : alk.
  paper) | ISBN 0268100535 (pbk. : alk. paper)
Subjects: LCSH: Death.
Classification: LCC HQ1073 .C88 2019 | DDC 306.9—dc23
LC record available at https://lccn.loc.gov/2019011956

∞ *This book is printed on acid-free paper.*

I dedicate this work to my husband, Dr. Lewis M. Cutter Jr.

While it may seem odd to dedicate a work on death

to my beloved spouse, think about it this way:

living well and dying well go hand in hand.

Lew, this book is dedicated to you.

We live so well; may dying well be in our future.

— Contents —

PART III: THE CHOICE OF DEATH

PART IV: THE LESSONS OF DEATH

— Detailed Table of Contents —

— Preface —

Death: A Reader presents a collection of primary source readings on how death is understood from prominent global perspectives that span ancient to contemporary discussions in philosophical and sacred texts. The collection includes views of death as physical disintegration, psychological disintegration, reincarnation, resurrection, medical immortality, digital immortality, an existential phenomenon of life, bad or good, to be feared or not, to be grieved and how, and to be hastened or not in the case of suicide, treatment refusal, and physician-assisted suicide. Primary source readings are contextualized for the reader and then serve as a basis for a set of exercises that engage readers in reflections about conceptual as well as practical issues surrounding death.

Four features of this text distinguish it from other texts on death. First, it provides a culturally diverse selection of primary source readings from philosophy and religion on the nature of death. Along with the more traditional discussions of death, it provides discussions on emerging topics in death studies, namely, medical immortality and digital immortality. Second, the text compiles a rich selection of discussions on the value of death and focuses on the extent to which death is good or bad, to be feared or not, and to be grieved and how. Third, it presents some of the key ethical issues on death, notably suicide, treatment refusal, and physician-assisted suicide, through the lens of the nature of death, thus linking popular bioethical topics on death with discussions of the metaphysics and epistemology of death. Fourth, the text offers practical opportunities to reflect on the nature of death through exercises that involve planning one's funeral and completing one's advance directives, as well as engaging in other activities related to the end of life.

Although the readings are drawn from primary source material from philosophical and sacred texts, *Death: A Reader* does not attempt to offer any particular philosophical, theological, doctrinal, or authoritative analysis of death. The analyses of death presented here are primarily descriptive and rely on generally accepted interpretations and translations of well-recognized writings in order to present an overview of a wide variety of worldviews on the nature of death. Although many important details about death in particular secular and sacred traditions are not provided here, readers are encouraged to spend more time studying the traditions that interest them. Further, the text does not delve deeply into some of the traditional problems that philosophers and religious studies scholars may expect, such as the mind-body problem and the nature of the soul, although these topics certainly arise in the readings. Should these topics be of interest to readers, there is a wealth of writings available online and through libraries. Again, readers are encouraged to pursue topics that interest them. Still further, the text does not seek to advance any particular view of death. It rather seeks to map out a geography of views of death for purposes of illustrating the wide variety of perspectives embraced by humans and offering an opportunity for readers to think through their own view of death. Finally, exercises that appear in the text are not meant to replace legal advice regarding end-of-life planning. Legal advice is readily available in most communities, although readers are reminded that a number of planning documents can be completed without the assistance of attorneys. Reflection exercises are designed for purposes of giving readers an opportunity to think about death and dying in a practical way.

Death: A Reader can be used in a variety of ways. In the classroom, whether this is live or online, it can be used to present both philosophical and sacred views of the nature of death in a philosophy or religious studies course. The text can also be used as a supplement in a course on death and dying in gerontology, psychology, and sociology and in nursing and medical schools to show the wide variety of views of death that health care professionals may encounter in the clinical setting. Alone, the text can be used in a book club or as a basis for personal reflection about what one thinks about death, one's own death as well as the death of another.

I come to this project after many years thinking about death and dying. In high school, I remember thinking at length about the choice Hamlet entertains, "To be or not to be." In college, I first read Søren Kierkegaard's *Fear and Trembling and the Sickness unto Death* and fell in love with philosophy and the nuanced interpretations it offers about seemingly settled matters. During my graduate studies in philosophy, my mother passed away, and I know that this loss framed my life and work in the years that came after. In my career, I have spent my academic life teaching a course on biomedical ethics spanning topics from birth to death. To date, I, along with everyone else, have experienced loss of loved ones, cherished possessions, hopes, abilities, and time on this earth. My interest in death and dying has not diminished; it has grown stronger as I see the relevancy of philosophical thinking on matters of loss and change.

This text could not have been written without support and advice from others. Each semester over the last many years, my undergraduate students at the University of Colorado at Colorado Springs (UCCS) have advised me about what materials speak to them and what reflection exercises assist them in thinking about death and dying. I thank my students in "Philosophical Issues in Death and Dying" at UCCS for their willingness to share their thoughts and suggestions. In addition, the text could not have been written without the support of my colleagues in the Department of Philosophy and administrators at UCCS who have continued to support my work in applied ethics and philosophy of medicine. I am particularly grateful to John and Vera Gardell, Anne Sarno, Tris Engelhardt, Raphael Sassower, Fred Bender, Lorraine Arangno, Rex Welshon, Sonja Tanner, and Jeff Scholes for their support of this project. Finally, I am indebted to my children, Lewis, Paige, Theresa, John, and Christine, who have supported my work in academics even when I may have been less than available at times. I am also indebted to my husband, Lewis M. Cutter Jr., who has unconditionally supported my work at the desk, in the classroom, and at home for many years now. As an academic, teacher, mother, and grandmother, I am deeply grateful for such support.

I continue to welcome comments about what could be changed or added to the readings and reflection exercises. Feel free to send comments to me at mcutter@uccs.edu. I can assure readers that your comments are appreciated.

July 18, 2018

— Chapter 1 —

Thinking about Death

A. WHY THINK ABOUT DEATH?

There are many reasons to think about death. To begin with, death is a centralizing notion and tells us something about what we share. While it is possible that a human might not have been born, once born, and at least so far in time, each and every human dies. Because death is shared, the study of death is bound to tell us something about what it means to be human. It is bound to tell us something about how we live our lives, treat the dying, organize our death rituals, expend resources in the dying process, formulate public policies and laws about death and dying practices, and make judgments about the praiseworthiness and blameworthiness of human actions in the context of our death and dying practices. Beyond this, and because of the inevitability of one's own death, a study of death is an opportunity for self-exploration and reflection upon what it means to be human, how we know and judge it, and how we think we ought to act in the face of such knowledge and judgments.

In addition, a study of death tells us something about how we live our lives differently. Despite the fact that we all die, humans do not share the same view of death. In ancient Greek philosophy, we find a number of accounts of death, including that death is loss of bodily heat. With developments in Asian and Abrahamic worldviews, death becomes an entryway to ultimate existence, *nirvana* or *brahman*, or paradise or hell. On such views, death involves loss of the soul and marks a point after which life continues, albeit in different ways, as illustrated in varying accounts of reincarnation and resurrection. With the rise of science and medicine in the late modern period, death is understood as a physical event or process involving biological disintegration, as in the case of defining death in terms of brain death. With the rise of the discipline of psychology, death is understood as a psychological process of disintegration, as in the case of defining death in terms of loss of mental capacities and properties such as consciousness and personal identity. In contemporary secular worldviews, death represents an opportunity for medical or digital immortality or a unique event in *my* life—views that challenge the universality of death as they emphasize the individuality and uniqueness of death.

How one views death frames how one values death. Put philosophically, a metaphysics and epistemology of death provides lenses through which evaluations of death are made. Death represents something that is bad or good, feared or not, to be grieved or not, and hastened or not, depending on one's view of death and the judgments of praiseworthiness or blameworthiness about death that are made. If, for instance, one views death as liberating the mind from the body, as the ancient Greek philosopher Plato does, then death is good. Alternatively, if one views death as the loss of a future life, as the contemporary American philosopher Thomas Nagel does, then death is bad. If, for instance, one holds that an individual has a right to die, as the Oregon Death with Dignity Act does, then death should be able to be hastened if the individual wishes it to be. Alternatively, if one holds that an individual does not have a right to die, as the Criminal Law of the People's Republic of China does, then death should not be able to be hastened. In these examples, how one views death provides a basis for how one values death. Given this, a study of death is an opportunity to explore ways to think about death, even when such ways differ from one's own personal view. It is an opportunity to explore and perhaps rethink how one thinks about death. Thus thinking about death is as much a personal as an academic exercise of reflecting on a shared human experience.

B. WHY STUDY GLOBAL PERSPECTIVES ON DEATH?

Death: A Reader brings together primary source materials on the nature of death drawn from a wide variety of philosophical and sacred worldviews. It draws from philosophical and sacred texts because world philosophies and religions have together helped us define and clarify the nature of death. Here *philosophical texts* refers to writings that formulate arguments devoted to the examination of basic concepts, such as death and life. For instance, Plato's *Republic* is heralded as a classic Western philosophical treatment of social and political life. *Sacred texts* refers to writings that are of central importance to a religious or spiritual tradition and are used in ritual, worship, or teaching settings. For instance, the Qur'an is a sacred text in the Abrahamic tradition of Islam revered by millions around the world. As a brief comparison, philosophical texts tend to use a rational methodology, and sacred texts tend to rely more on articles of faith in their investigations. One thinks here of German philosopher Martin Heidegger's argument in *Being and Time* about the nature of death as the possibility of the absolute impossibility of the *Da-Sein* and Matthew's testimony of death in the New Testament based on his belief in Jesus Christ. Yet numerous philosophical and sacred texts defy strict categorization and even bleed into one another. One thinks here of the Bhagavad Gita, which is considered one of the sacred texts of Hinduism *and* one of the most influential philosophical works in human history. In the end, we can say that philosophy and the world religions share an interest in fundamental questions concerning death and life and offer a wide range of sources for our academic and personal reflections.

In the chapters that follow, we explore this shared interest in thinking about death through diverse frameworks. Presented in a time line, the selected and/or discussed thinkers and religious figures are:

Before Common Era (BCE)	*Author and/or Work (Place of Writing)*
ca. 8000 BCE	Australian Dreamtime stories (Australia)
ca. 2000	*Epic of Gilgamesh* (Mesopotamia)
ca. 1580–1090	*Egyptian Book of the Dead* (Egypt)
ca. 1500–600	Vedas (India)
ca. 1400–1200	Zarathustra (ancient Greater Iran)
ca. 1300	*Papyrus of Nu*, in *The Egyptian Book of the Dead* (Egypt)
ca. 850	Homer (Greece)
8th c.	*Avesta: Yasna* (ancient Greater Iran)
ca. 800–300	Upanishads (India)
ca. 6th to 4th c.	Buddha (Nepal)
ca. 580	Book of Ezekiel, in the Hebrew Bible (Middle East)
ca. 575	Lao-Tzu, *Tao Te Ching* (China)
ca. 551–497	Confucius (China)
ca. 500	*Sutta Pitaka* of the *Tripitaka* (India)
ca. 470–399	Socrates (Greece)
ca. 427–348	Plato (Greece)
ca. 400	Bhagavad Gita (India)

after 5th c.	*Katha Upanishad* (India)
384–322	Aristotle (Greece)
ca. 372–289	Mencius (China)
ca. 369–286	Chuang Tzu (China)
341–271	Epicurus (Greece)
ca. 3rd c.	*Dhammapada* (Nepal)
ca. 2nd c.	Book of Daniel, in the Hebrew Bible (Middle East)
106–43	Cicero (Italy)
ca. 6 BCE–30 CE	Jesus Christ (Galilee, Roman Judea)
ca. 4 BCE–65 CE	Seneca (Italy)

Common Era (CE)

ca. 50–ca. 130	Epictetus (Greece)
121–80	Marcus Aurelius (Italy)
end 1st c.	Book of Matthew, in the New Testament (Galilee or Judea)
354–430	Augustine (Algeria)
610–32	Muhammad (Mecca, Saudi Arabia)
7th c.	*Noble Qur'an* (Saudi Arabia)
8th c. (ca. 712)	*Kojiki* (Japan)
ca. 8th c.	*Tibetan Book of the Dead* (Tibet)
1225–74	Thomas Aquinas (Italy)
1438–1471/1472	Pachacuti Inca Yupanqui (Peru)
1588	Michel de Montaigne (France)
1649	René Descartes (France)
1689	John Locke (England)
early 1700s	Yamamoto Tsunetomo, *Hagakure* (Japan)
1755	David Hume (Scotland)
1785	Immanuel Kant (Germany)
1849	Søren Kierkegaard (Denmark)
1927	Martin Heidegger (Germany)
1943	Jean-Paul Sartre (France)
1960	Birago Diop (Mali)
1961	C. S. Lewis (England)
1964	Simone de Beauvoir (France)

1969	Elisabeth Kübler-Ross (Switzerland)
1973	Ernest Becker (Canada)
1974	Abraham Joshua Heschel (New York)
1979	Thomas Nagel (New York)
1981	President's Commission (Washington, D.C.)
1985	Simone de Beauvoir (France)
1993	Robert M. Veatch (Washington, D.C.)
1994	National Association of Evangelicals (United States)
1997	Ben A. Rich (California)
1997	*Catechism of the Catholic Church* (Washington, D.C.)
2000	John Harris (England)
2001	Rimpoche Gehlek Nawang (Tibet)
2002	Abraham Rudnick (Israel)
2002	Eelco F. M. Wijdicks (Minnesota)
2002	Netherlands Termination of Life on Request and Assisted Suicide (Review Procedures) Act (Netherlands)
2003	Tom Pyszczynski (Colorado), Sheldon Solomon (New York), and Jeff Greenberg (Arizona)
2005	Jung Kwak (Wisconsin) and William Haley (Florida)
2008	George J. Annas (Massachusetts)
2009	Carol Forsloff (Louisiana)
2009	President's Council on Bioethics (Washington, D.C.)
2010	Kristie McNealy (England)
2010	Anthony N. DeMaria (California)
2011	Chris Faraone (Massachusetts)
2011	Wendy Zeldin (Washington, D.C.)
2012	Patrick Stokes (Australia)
2012	Susan Steiner (England)
2012	Carla Sofka (New York)
2012	Leung Wing Chu (Hong Kong)
2013	Kelly Allen (Tennessee)
2013	Colorado Medical Treatment Act (Colorado)
2013	Colorado Medical Durable Power of Attorney (Colorado)
2013	Criminal Law of the People's Republic of China (China)

| 2013 | Oregon Death with Dignity Act (Oregon) |
| 2016 | Richard A. Jones (Washington, D.C.) |

Selections span a period from approximately 8000 BCE to the present and cover a range of topics and views drawn from philosophical and sacred writings from the East and West. They include writings from Africa, the Middle East, Europe, India, China, Japan, Australia, South America, and North America to present a global perspective of death. While these categories of temporal and geographical location are useful, readers are reminded that views of death within particular cultural and spiritual traditions typically defy simple classification and description. Stories that emerge in history and span generations, such as the stories about death from African and Native American tribes, are not easily located by a simple date and configured in a simple time line. Geographical categories, such as "Asian," "Eastern," and "Western," are also not so simple; they are to be taken as ways to cluster traditions of thought that share some notable similarities. Readers are encouraged to keep in mind that the categories we employ are neither simple nor unchanging.

C. OUTLINE OF TOPICS

What follows are prominent themes that emerge and reemerge in the history of ideas regarding the nature of death. Included here are short selections of primary source readings from a variety of traditions—Chinese, Indian, ancient Greek and Roman, African, Native American, Tibetan Buddhist, Taoist, Shinto, Aboriginal Australian, and Abrahamic (Islamic, Judaic, and Christian)—and frameworks that include early Western skepticism, nineteenth- and early twentieth-century existentialism and phenomenology, late twentieth-century and early twenty-first-century psychology, contemporary digital philosophy, and contemporary public policy and legal approaches. Selections are short in order to expose readers to traditions of thought, but readers are encouraged to consult the entire book, article, or legislation should they have interest.

Four parts organize the readings in this book: "The Nature of Death," "The Value of Death," "The Choice of Death," and "The Lessons of Death." Fourteen themes are presented in order to convey a sense of major views of death on a global scale. Each theme is stated as an option for inclusion in the sentence "Death is _____." In part I, "The Nature of Death," death can be understood in terms of physical disintegration, psychological disintegration, reincarnation, resurrection, medical immortality, digital immortality, and an existential phenomenon of life. In part II, "The Value of Death," death can be understood in terms of bad or good, to be feared or not, and to be grieved and how. In part III, "The Choice of Death," death can be understood as something that can be hastened by suicide, treatment refusal, or physician-assisted suicide. In part IV, "The Lessons of Death," death can be understood in terms of the instruction that it imparts. Each theme is organized in terms of the context of the discussion, perspectives offered through primary source readings, reflection exercises, and suggestions for further reading.

More specifically, each theme is introduced with a brief background discussion of the historical and conceptual context of the view that is presented. Each theme is divided into subthemes to provide a sense of their varying expressions. Each subtheme is also stated as an option for inclusion in the sentence "Death is _____." Typically, the series of short readings from primary sources are ordered historically within each section and span a page or more of reading. Reading selections are primarily from philosophical and sacred sources, although some perspectives are drawn from medicine, public policy, and the law and convey notable positions on the nature of death influenced by philosophical or sacred traditions of thought. Reading selections are followed with opportunities for reflection on the nature and meaning of death. Some are analysis reflections on the readings, and others are practical. Analysis questions are marked by an asterisk (*) and can be answered by considering the readings and going to outside sources to help interpret the readings. In my own classroom, I ask students to write a page-long answer to questions with an asterisk to ensure a minimum level of development of thought in the response. Practical exercises supplement academic reflections. Practical exercises are drawn from US practices; readers from other countries are encouraged to find out if there are analogous end-of-life planning documents in their own countries. The exercises include planning one's funeral or memorial service, making organ donation requests, pricing out the cost of one's funeral or memorial service, observing Asian and Western funerals or memorial services, pricing out the cost of dying, writing one's obituary, completing a bucket list, organizing one's important documents and contacts, writing a last letter, assessing one's fear of death and how one grieves, completing one's last will and testament, finalizing one's living will and medical durable power of attorney, and requesting physician-assisted suicide. For the practical exercises, access to online materials is sometimes required. Exercises are in no way meant to force a reader into accepting a certain death and dying practice. Exercises are developed to expose readers to a variety of views of death and to encourage readers to think through in a personal way their own commitments. Each chapter ends with a list of suggestions for further reading. Together, the ensemble of views about death and opportunities for reflection indicates that there is no single story about death.

D. REFLECTIONS: FUNERAL OR
MEMORIAL SERVICE PLANNING

1. A beginning reflection: summarize who you are and what is important to you in life.

2. Reflect on your views about death and dying:
    a. Why are you interested in studying about death?
    b. What, in your view, is death?
    c. What, in your view, is the value of death?
    d. Do you think that we, as humans, have a choice regarding death and how we die?
    e. Feel free to share anything else about your view of death and dying.

3. Plan your own funeral or memorial service. Have you made any plans for your funeral or memorial service?
    a. If so, briefly summarize what you have planned.
    b. If not, plan your funeral or memorial service. Here are some questions to get started:
        i. Would you like to be buried or cremated?
        ii. If buried, do you want to be buried in a cemetery, in a churchyard, in a vault, or elsewhere? Do you wish a special coffin or a "green burial" (i.e., an environmentally conscious one)? Do you wish to be embalmed, to be buried in a shroud, to wear a special outfit, or to be buried with special possessions? If so, give details.
        iii. If cremated, do you wish a special urn, or special possessions placed in the urn? Do you intend on having someone keep your ashes or will they be scattered? If someone will keep you, who will this person be? If your ashes will be scattered, where do you want that to happen? Give details.
        iv. Would you like to have a church or temple service, a religious service not in a church or temple, a graveyard service, a nonreligious service, or no service? Would you like a private or public service? Give details.
        v. Would you like your body to be repatriated (transferred to another country)? If so, give details.
        vi. Would you like special readings, music, flowers, and/or a eulogy (i.e., an oral speech given by a loved one in celebration of your life)? If so, give details.
        vii. Would you like to have an obituary appear in a local, regional, national, or international newspaper or social website? If so, give details.

viii. Would you like to give your loved ones an opportunity to make donations to a special charity? If so, give details.

ix. Would you like a special gathering of friends and family after the funeral service? If so, specify who is invited, what food and drinks you would like, and any other special activities. Give details.

x. Would you like a lasting memorial, such as a bench, a tree, your name on a building, or a charity fund in your name? If so, give details.

xi. Share anything else you would like about your funeral or memorial service.

E. GENERAL REFERENCE MATERIAL

Numerous texts, websites, and films feature opportunities to think about death. What follows is a selection of general reference materials readers may find helpful in providing further background on the nature of death.

1. Books and Articles

Andrade, Gabriel. 2017. "Immortality." In *Internet Encyclopedia of Philosophy.* https://www.iep.utm.edu/immortal/.

Aries, Philippe. 1975. *Western Attitudes toward Death: From the Middle Ages to the Present.* Baltimore: Johns Hopkins University Press.

Barry, Vincent. 2007. *Philosophical Thinking about Death and Dying.* Belmont, CA: Thomson Wadsworth.

Bilhartz, Terry D. 2006. *Sacred Words: A Source Book on the Great Religions of the World.* New York: McGraw Hill.

Brennan, Samantha, and Robert J. Stainton, eds. 2009. *Philosophy and Death: Introductory Readings.* Toronto, Ontario: Broadview Press.

Capps, Walter H. 1995. *Religious Studies: The Making of a Discipline.* Minneapolis: Augsburg Fortress.

Carr, Thomas K. 2006. *Introducing Death and Dying: Readings and Exercises.* Upper Saddle River, NJ: Pearson/Prentice-Hall.

Chidester, David. 2002. *Patterns of Transcendence: Religion, Death, and Dying.* 2nd ed. Belmont, CA: Wadsworth.

Critchley, Simon. 2008. *The Book of Dead Philosophers.* New York: Vintage Books.

DeGrazia, David. 2017. "The Definition of Death." In *Stanford Encyclopedia of Philosophy.* http://plato.stanford.edu/entries/death-definition/.

Despelder, Lynne Ann, and Albert Lee Strickland. 2014. *The Last Dance: Encountering Death.* 10th ed. New York: McGraw Hill.

Edwards, Paul, ed. 1997. *Immortality.* New York: Prometheus Books.

Encyclopedia of Death and Dying. 2017. www.deathreference.com.

Feifel, Herman. 1959. *The Meaning of Death.* New York: McGraw-Hill.

Fingarette, Herbert. 1996. *Death: Philosophical Soundings.* Chicago: Open Court.

Foley, Elizabeth. 2011. *The Law of Life and Death.* Cambridge, MA: Harvard University Press.

Hasker, William, and Charles Taliaferro. 2017. "Afterlife." In *Stanford Encyclopedia of Philosophy.* http://plato.stanford.edu/entries/afterlife/.

Hayasaki, Erika. 2014. *The Death Class: A True Story about Life.* New York: Simon and Schuster.

Kagan, Shelly. 2012. *Death.* New Haven, CT: Yale University Press.

Kastenbaum, Robert. 2016. *Death, Society, and Human Experience.* 11th ed. New York: Pearson Education.

Kramer, Kenneth. 1988. *The Sacred Art of Dying.* Mahwah, NJ: Paulist Press.

Kramer, Scott, and Kuang-Mong Wu. 1988. *Thinking through Death.* 2 vols. Malabar, FL: Robert E. Krieger.

Kushner, Harold. 1978. *When Bad Things Happen to Good People.* New York: Random House.

Lizza, John. 2006. *Persons, Humanity, and the Definition of Death.* Baltimore: Johns Hopkins University Press.

Luper, Steven. 2009. *The Philosophy of Death.* Cambridge: Cambridge University Press.

———. 2017. "Death." In *Stanford Encyclopedia of Philosophy.* http://plato.stanford.edu/entries/death/.

Malpas, Jeff E., and Robert C. Solomon. 1998. *Death and Philosophy.* London: Routledge.

McQuire, Meredith. 2008. *Religion: The Social Context.* 5th ed. New York: Waveland Press.

Neuhaus, Richard John. 2000. *The Eternal Pity.* Notre Dame: University of Notre Dame Press.

Nuland, Sherwin. 1995. *How We Die: Reflections on Life's Final Chapter.* New York: Vintage.

Pallis, Christopher A. 2017. "Death." *Encyclopedia Britannica.* www.britannica.com/EBchecked/topic/154412/death.

Pojman, Louis P. 2002. *Life and Death: Grappling with the Moral Dilemmas of Our Times.* Belmont, CA: Wadsworth.

Rosenberg, Alexander. 1998. *Thinking Clearly about Death.* 2nd ed. Indianapolis: Hackett.

Singer, Peter. 1994. *Rethinking Life and Death.* New York: St. Martin's Press.

Smith, Huston. 1991. *The World's Religions.* New York: HarperSanFrancisco.

Sommers, Deborah, ed. 1995. *Chinese Religion: An Anthology of Sources.* New York: Oxford University Press.

2. Films

Being Mortal (Frontline Productions, 2015): a documentary that explores the intersection of life, death, and medicine.

Bucket List (2007): the fictional story of the bond between two cancer patients who decide that they are not going to accept their fate.

Dead Man Walking (1995): drama based on the memoir of a nun, Helen Prejean, who helps the family of a murdered girl and the man who is on death row for the crime.

Facing Death (Frontline Productions, PBS, 2010): a documentary set at the Mount Sinai Medical Center in New York about today's complicated end-of-life decisions.

Fault Lines—Dying Inside: Elderly in Prison (Al Jazeera English, 2010): an investigation of how the US prison system handles an increasing number of aging prisoners.

How to Die in Oregon (HBO Documentary Films, 2011): a look at Oregon's legalization of physician-assisted suicide.

In the Matter of Karen Ann Quinlan (1977): the true story of a young adult whose parents were faced with decisions about what should be done at the end of her life.

Life before Death (Moonshine Movies, 2012): an exploration of how health care professionals from eleven countries manage the pain and suffering of patients before their death.

Moment of Death (National Geographic, 2008): a documentary that looks at the physical and psychological changes experienced by a human in the moments before and after death.

The Space Between (Passion Projects, 2013): a look at a nurse practitioner's efforts to partner with local leaders in Kenya to open a hospice and provide palliative care to the dying.

The Suicide Plan (Frontline Productions, 1983): a look at the underground support for physician-assisted suicide in the United States.

Tuesdays with Morrie (Touchstone Entertainment, 1999): a retelling of conversations between a student and his dying professor.

What Dreams May Come (1998): a fictional story about what happens after death.

Whose Life Is It Anyway? (1981): a fictional story about a paralyzed artist who seeks court support to be allowed to die.

You Don't Know Jack (2010): a biopic on the physician, Dr. Jack Kevorkian, who carried out physician-assisted suicide and advocated on behalf of its legalization in the 1990s.

Part I

THE NATURE
OF DEATH

Part I approaches the nature of death from a variety of perspectives. In philosophy and religious studies, the study of the nature of a phenomenon is found within the academic field called *ontology* (from the Greek *onto-*, meaning "being," and *-ology*, meaning "study of"). This field is rooted in ancient quests to understand the nature of reality. An ontological view of death, for instance, might hold that death is the loss of a transcendent entity such as a soul or might hold that death is nothing more than the loss of physical integration of the body. Complementing an ontological account of reality is an epistemological one. The study of how we know a phenomenon or reality is found within the academic field called *epistemology* (from the Greek *episteme,* meaning "knowledge"). This field is rooted in ancient quests to understand how we know what we think we know. For example, the ontological view that death is loss of that which is transcendent is typically coupled with some type of epistemological approach that verifies the loss of that which is transcendent. For instance, the view that death occurs with the loss of the soul is typically supported by some argument about what the loss of the soul entails. The view that death is loss of physical integration of the body is typically coupled with some type of epistemological approach that verifies the loss of physical integration. For instance, the view that death is a biological event is typically supported by empirical measures of the loss of bodily function, such as those made possible by an electroencephalogram (EEG).

The readings in this first part explore a range of ontological and epistemological views of death drawn from philosophical and sacred texts. These include a wide variety of positions, including death as physical disintegration, psychological disintegration, reincarnation, resurrection, medical immortality, digital immortality, and an existential phenomenon of life. The positions here are drawn from a variety of traditions, including old and new, and East and West. In the readings, consider the claims and assumptions made in each of the perspectives on death. In doing so, readers may notice some notable shared as well as divergent views on the nature of death. Correspondingly, readers may find that they agree with some of the authors' claims and disagree with others. Readers are encouraged to note these points of agreement and disagreement in preparation for the exercises that appear at the end of each chapter.

Chapter 2

Physical Disintegration

A. CONTEXT

It is not difficult in contemporary culture to find the view that death is what I call "physical disintegration." This position ranks as one of the more prominent accounts found in the twentieth and twenty-first centuries because of support provided by science and medicine for the view that death is loss of biological function. Science and medicine give us widely accepted empirical support for the view that death involves loss of matter, loss of reflexes, lack of response to vigorous external stimuli, and absence of electrical activity as evidenced by a flat electroencephalogram (EEG).

The US Department of Health and Human Services, in conjunction with the US Centers for Disease Control, produces an annual report on the leading causes of death in the United States. A 2012 report presents 2009 data on the ten leading causes of death in the United States. For individuals aged sixty-five and over, they are (along with the total number of deaths):

(1)   diseases of the heart (599,413)
(2)   malignant neoplasms (cancers) (567,413)
(3)   chronic lower respiratory diseases (137,353)
(4)   cerebrovascular diseases (128,842)
(5)   accidents (unintentional injuries) (118,021)
(6)   Alzheimer's disease (79,003)
(7)   diabetes mellitus (68,705)

(8)   influenza and pneumonia (53,692)

(9)   nephritis, nephrotic syndrome, and nephrosis (48,935)

(10)   intentional self-harm (suicide) (36,909)

(Heron 2012, 9)

For those ages fifteen to twenty-four in the United States, the leading causes of death are accidents, suicide, homicide, and cancer ("What Are My Risks Factors?" 2015, 70–71).

Worldwide, ischemic heart disease, stroke, chronic obstructive pulmonary disease, lower respiratory infections, Alzheimer's disease and other dementias, trachea/bronchus/lung cancers, and diabetes have remained the top major causes of death in the first decade of the twenty-first century (World Health Organization 2018). Diarrhea remains one of the top ten causes of death worldwide, leading to 1.6 million deaths in 2016. Road injuries leading to death remain a significant problem on the global scale. Deaths from HIV infection have decreased, from 1.5 million to 1.0 million between 2000 and 2016 (World Health Organization 2018).

The view that death is physical disintegration has been promulgated in a report on death written by the President's Commission for the Study of Ethical Problems in Medicine and Biomedical and Behavioral Research (1981). Today, this report serves as the basis of state laws defining death in the United States. Similar standards can be found in other countries as well, as physician Eelco Wijdicks (2002) reports, although there are notable differences among the diagnostic criteria for death worldwide. A more recent president's commission, the President's Council on Bioethics (2009), takes up debates about the neurological determination of death in a section entitled "Controversies on the Determination of Death."

The view that death is a biological event is not simply a contemporary one. One can find its roots in the writings of the ancient Greek philosopher Aristotle and the early modern French philosopher René Descartes. Although Aristotle and Descartes are neither strict materialists nor empiricists, they each express views that death can be defined in terms of the loss of physical matter. Here, a *materialist* (from the Latin *materia*, meaning "matter") holds the ontological view that physical or material matter is the fundamental substance in nature. In the case of death, and on a materialist view, death is a physical event or process that can be explained in terms of loss of physical integration. An *empiricist* (from the Greek *empeiria*, meaning "experience") holds the epistemological view that knowledge is ultimately derived from the senses, or experience, and that ideas can be traced to sense data. In the case of death, and on an empiricist view, death is a physical event or process composed of interactions among matter. Such a view is established by sense experience or observation made possible by biomedical technology. The selections that follow illustrate the widely popular view in the twentieth and twenty-first centuries that death is the loss of physical integration of the body.

B. PERSPECTIVES

1. Lack of Heat

Ancient Greek philosopher Aristotle (384–322 BCE) is known for his contributions to logic, metaphysics, mathematics, physics, biology, botany, ethics, politics, and medicine. He was a student of the ancient Greek philosopher Plato (ca. 427–348 BCE) (for more on Plato, see chapter 4, B.2.a) and investigated many of the most enduring questions in philosophy concerning reality, knowledge, and how to live. Aristotle is particularly known for his work *Nicomachean Ethics*, which provides an early systematic account of how to live the ethical life through a life of virtue. The word *virtue* comes from the Latin *virtus*, meaning "moral perfection" or "valor," and refers to an appropriately expressed character trait. Aristotle is also known for his view that reality is composed of a nonmaterial form. Such reality (including matter) is known through rational as well as empirical observations. In this way, Aristotle is neither a strict materialist nor an empiricist, although he provides a partial physical explanation for death ("Aristotle" 2017).

For Aristotle, death is the loss of "breath" or respiration and comes about through a process by which the body loses heat and grows cold and dry. As he says, "Death is the departure of those forms owing to the expulsive pressure exerted by the surrounding air" (1984, 751 [472a11–17]). In line with his interest in classifying things in nature, Aristotle distinguishes among types of death, including violent and natural death. Violent death occurs "when the cause of death is external" (760 [478b22]), as in the case of death by murder. With regard to natural death, Aristotle tells us that "it is always to some lack of heat that death is due, and in perfect creatures the cause is its failure in the organ containing the source of the creature's essential nature" (760 [478b32–34]). Here *heat* refers to the natural organic force that gives life and motion to the body. In this way, heat is both matter and force and involves material as well as nonmaterial components. What is of interest here is Aristotle's observation that death is loss of heat. Such a view coincides with clinical observations today. In the following selection from *On Youth, Old Age, Life and Death, and Respiration*, Aristotle provides an early version of a biological account of death.

Reading: *On Youth, Old Age, Life and Death, and Respiration* (Aristotle 1984, 751 [472a11–17], 760–61 [478b21–479a27]). Reprinted with the permission of Princeton University Press.

[p. 751 (472a11–17)]

This explains why life and death are bound up with the taking in and letting out of the breath; for death occurs when the compression by the surrounding air gains the upper hand, and, the animal being unable to respire, the air from out-side can no longer enter and counteract the compression. Death is the depar-ture of those forms owing to the expulsive pressure exerted by the surrounding air. As to the reason why all must die at some time—not, however, at any chance time but, when natural, owing to old age, and, when unnatural, to violence. . . .

[pp. 760–61 (478b21–479a27)]

To be born and die is common to all animals, but there are specifically di-verse ways in which these phenomena occur; of destruction there are different types, though yet something is common to them all. There is violent death and again natural death, and the former occurs when the cause of death is external, the latter when it is internal, and involved from the beginning in the constitution of the organ, and not an affection derived from a foreign source. In the case of plants, the name given to this is withering, in animals old age. Death and decay pertain to all things that are not imperfectly developed; to the imperfect also they may be ascribed in nearly the same but not an identical sense. Under the imper-fect I call eggs and seeds of plants as they are before the root appears.

It is always to some lack of heat that death is due, and in perfect creatures the cause is its failure in the organ containing the source of the creature's essen-tial nature. This member is sited, as has been said, at the junction of the upper and lower parts; in plants it is intermediate between the root and the stem, in san-guineous animals it is the heart, and in those that are bloodless the correspond-ing part of their body. But some of these animals have potentially many sources of life, though in actuality they possess only one. This is why some insects live when divided, and why, even among sanguineous animals, all whose vitality is not intense live for a long time after the heart has been removed. Tortoises, for example, do so and make movements with their feet, so long as the shell is left, a fact to be explained by the natural inferiority of their constitution, as it is in in-sects also.

The source of life is lost to its possessors when the heat with which it is bound up is no longer tempered by cooling, for as I have often remarked, it is consumed by itself. Hence when, owing to lapse of time, the lung in one class and the gills in the other get dried up, these organs become hard and earthy and incapable of movement and cannot be expanded or contracted. Finally things come to a climax, and the fire goes out from exhaustion.

Hence a small disturbance will speedily cause death in old age. Little heat remains, for the most of it has been breathed away in the long period of life preceding, and hence any increase of strain on the organ quickly causes extinction. It is just as though the heart contained a tiny feeble flame which the slightest movement puts out. Hence in old age death is painless, for no violent disturbance is required to cause death, and the severance of the soul is entirely imperceptible. All diseases which harden the lung by forming tumours or waste residues, or by excess of morbid heat, as happens in fevers, accelerate the breathing owing to the inability of the lung to move far either upwards or downwards. Finally, when motion is no longer possible, the breath is given out and death ensues.

2. Loss of Bodily Movement

Seventeenth-century French philosopher René Descartes (1596–1650) is known for his contributions to epistemology and our thinking about knowledge in the early years of the development of science. Notable is his defense of the claim that "I am, [therefore] I exist" (Descartes [1641] 2009, 139), which remains a major focus of study in contemporary philosophy. This view, often stated as "I think therefore I am," comes to be a rallying call for modern *rationalists*, who support the view that reason provides access to knowledge. In addition, and particularly in the *Discourse on Method*, Descartes seeks to ground science and medicine on foundations that are clear and distinct, as opposed to speculative and opinion based. He recommends what we today call the "scientific method," which involves close observation, accumulation of data, and interpretation of data in order to support or reject a hypothesis (Descartes [1641] 2009, 31). Here Descartes's commitment to an empirical methodology of knowing is evident: we know by observing our world. Yet, like Aristotle, Descartes is neither a strict materialist nor an empiricist. He holds that a human is composed of body as well as mind (or what he calls soul). Descartes's "dualist" view of a human being has been notably influential in the West, leading to ways in which we see and structure the world in terms of empirical and nonempirical entities.

In the following selection from *Passions of the Soul*, Descartes provides an early modern version of a biological view of death. Descartes argues that in death the soul, which is understood as a nonempirical or noncorporeal substance, "quits" the body when the "principal parts" decay and lose both heat and motion. As he says, "The soul takes its leave when we die *only because* this heat ceases and the organs which bring about bodily movement decay" (Descartes [1649] 1985, 329, my italics). Here, heat is "the corporeal [bodily] principle of the movements for which it is designed" (Descartes [1649] 1985, 330). It is responsible for life and makes possible bodily movement. Alternatively, loss of heat leads to loss of bodily movement and the loss of the soul, which brings about death.

Reading: *Passions of the Soul* (Descartes [1649] 1985, 328–30 [arts. 2–6]). Reprinted with the permission of Cambridge University Press.

[Article 2]
To understand the passions of the soul we must distinguish its functions from those of the body
Next I note that we are not aware of any subject which acts more directly upon our soul than the body to which it is joined. Consequently we should recognize that what is a passion in the soul is usually an action in the body. Hence there is no better way of coming to know about our passions than by examining the difference between the soul and the body, in order to learn to which of the two we should attribute each one of the functions present in us.

[Article 3]
The rule we must follow in order to do this
We shall not find this difficult if we bear in mind that anything we experience as being in us, and which we see can also exist in wholly inanimate bodies, must be attributed only to our body. On the other hand, anything in us which we cannot conceive in any way as capable of belonging to a body must be attributed to our soul.

[Article 4]
The heat and the movement of the limbs proceed from the body, and the thoughts from the soul
Thus, because we have no conception of the body as thinking in any way at all, we have reason to believe that every kind of thought present in us belongs to the soul. And since we do not doubt that there are inanimate bodies which can move in as many different ways as our bodies, if not more, and which have as much heat or more (as experience shows in the case of a flame, which has in itself much more heat and movement than any of our limbs), we must believe that all

the heat and all the movements present in us, in so far as they do not depend on thought, belong solely to the body.

[Article 5]
It is an error to believe that the soul gives movement and heat to the body
In this way we shall avoid a very serious error which many have fallen into, and which I regard as the primary cause of our failure up to now to give a satisfactory explanation of the passions and of everything else belonging to the soul. The error consists in supposing that since dead bodies are devoid of heat and movement, it is the absence of the soul which causes this cessation of movement and heat. Thus it has been believed, without justification, that our natural heat and all the movements of our bodies depend on the soul; whereas we ought to hold, on the contrary, that the soul takes its leave when we die only because this heat ceases and the organs which bring about bodily movement decay.

[Article 6]
The difference between a living body and a dead body
So, as to avoid this error, let us note that death never occurs through the absence of the soul, but only because one of the principal parts of the body decays. And let us recognize that the difference between the body of a living man and that of a dead man is just the difference between, on the one hand, a watch or other automaton (that is, a self-moving machine) when it is wound up and contains in itself the corporeal principle of the movements for which it is designed, together with everything else required for its operation; and, on the other hand, the same watch or machine when it is broken and the principle of its movement ceases to be active.

3. Irreversible Cessation of All Functions of the Entire Brain

The President's Commission for the Study of Ethical Problems in Medicine and Biomedical and Behavioral Research was commissioned by the US Congress and met in Washington, D.C., between 1978 and 1983 to investigate medical, biomedical, and behavioral research activities that raise bioethical issues. *Bioethics* (from the Greek *bio*, meaning "life" and *ēthikē*, meaning "ethical") is the study of the ethical or moral implications of biomedical discoveries and practices. It gained recognition at the end of the twentieth century for its incisive analyses and critiques of practices in medicine. The term *bioethics* was coined by Dr. Van Rensselaer Potter, a research oncologist at the University of Wisconsin in the early 1970s (Jonsen 1998, 27). Potter published an article in 1970 entitled "Bioethics: The Science of Survival" and in 1971 followed it with his book *Bioethics: Bridge to the Future*. Potter defines "bioethics" as "a new discipline that combines biological knowledge with a knowledge of human value systems" (1971, 2). Bioethics, or biomedical ethics, has since become influential in Western medicine especially as concerns about the role, power, and limits of medicine in our lives abound.

The President's Commission generated a well-received empirical definition of brain death that has ultimately served as the basis of state law in the United States defining death. The definition, known as the Uniform Determination of Death Act, reads: "An individual who has sustained either (1) irreversible cessation of circulatory and respiratory functions, or (2) irreversible cessation of all functions of the entire brain, including the brain stem, is dead. A determination of death must be made in accordance with accepted medical standards" (President's Commission 1981, 2). The definition is the basis of the clinical signs of death, such as absence of pupil reaction to light, of jaw or gag reflex, and of an individual's reaction to pain over a period of time. Such signs indicate that all parts of the brain have ceased to function. These parts of the brain include the cerebrum (which controls higher-brain functions, such as thought and language), the cerebellum (which controls coordination of movement and balance), the limbic system (which controls feelings and emotions), and the brain stem (which regulates basic life functions, such as breathing and heartbeat). Signs of death are verified by a flat-line electroencephalogram [EEG] or brain wave assessment. The definition of death that is presented here represents a "whole-brain" or "whole-brain-oriented" definition of death and relies on empirical evidence made possible by clinical tests.

Reading: *Defining Death: Medical, Legal and Ethical Issues in the Determination of Death* (President's Commission 1981, 1–2 and 31–35)

[pp. 1–2]
The enabling legislation for the President's Commission directs it to study "the ethical and legal implications of the matter of defining death, including the advisability of developing a uniform definition of death" (42 U.S.C. §1802 [1978]). In performing its mandate, the Commission has reached conclusions on a series of questions which are the subject of this Report. In summary, the central conclusions are:

1. That recent developments in medical treatment necessitate a restatement of the standards traditionally recognized for determining that death has occurred.

2. That such a restatement ought preferably to be a matter of statutory law.

3. That such a statute ought to remain a matter for state law, with federal action at this time being limited to areas under current federal jurisdiction.

4. That the statutory law ought to be uniform among the several states.

5. That the "definition" contained in the statute ought to address general physiological standards rather than medical criteria and tests, which will change with advances in biomedical knowledge and refinements in technique.

6. That death is a unitary phenomenon which can be accurately demonstrated either on the traditional grounds of irreversible cessation of heart and lung functions or on the basis of irreversible loss of all functions of the entire brain.

7. That any statutory "definition" should be kept separate and distinct from provisions governing the donation of cadaver organs and from any legal rules on decisions to terminate life-sustaining treatment.

To embody these conclusions in statutory form the Commission worked with the three organizations which had proposed model legislation on the subject, the American Bar Association, the American Medical Association, and the National Conference of Commissioners on Uniform State Laws. These groups have now endorsed the following statute, in place of their previous proposals:

Uniform Determination of Death Act

An individual who has sustained either (1) irreversible cessation of circulatory and respiratory functions, or (2) irreversible cessation of all functions of the entire brain, including the brain stem, is dead. A determination of death must be made in accordance with accepted medical standards.

The Commission recommends the adoption of this statute in all jurisdictions in the United States. [*Ed. note: all states in the United States have since adopted this definition of death.*] . . .

[pp. 31–35]

Understanding the "Meaning" of Death

It now seems clear that a medical consensus about clinical practices and their scientific basis has emerged; certain states of brain activity and inactivity, together with their neurophysiological consequences, can be reliably detected and used to diagnose death. To the medical community, a sound basis exists for declaring death even in the presence of mechanically assisted "vital signs."

Yet before recommending that public policy reflect this medical consensus, the Commission wished to know whether the scientific viewpoint was consistent with the concepts of "being dead" or "death" as they are commonly understood in our society. These questions have been addressed by philosophers and theologians, who have provided several formulations.

The Commission believes that its policy conclusions . . . must accurately reflect the social meaning of death and not constitute a mere legal fiction. The Commission has not found it necessary to resolve all of the differences among the leading concepts of death because these views all yield interpretations consistent with the recommended statute.

Three major formulations of the meaning of death were presented to the Commission: one focused upon the functions of the whole brain, one upon the functions of the cerebral hemispheres, and one upon non-brain functions. Each of these formulations (and its variants) is presented and evaluated. [*Ed. note: The whole-brain definition of death, which is supported by the President's Commission, is presented here. An elaboration of the cerebral or higher-brain view is found in chapter 3 of this volume. Elaborations of "nonbrain" views of death are found in chapters 4, 5, and 7 of this volume.*]

The "Whole Brain" Formulations

One characteristic of living things which is absent in the dead is the body's capacity to organize and regulate itself. In animals, the neural apparatus is the dominant locus of these functions. In higher animals and man, regulation of both maintenance of the internal environment (homeostasis) and interaction with the external environment occurs primarily within the cranium.

External threats, such as heat or infection, or internal ones, such as liver failure or endogenous lung disease, can stress the body enough to overwhelm its ability to maintain organization and regulation. If the stress passes a certain level, the organism as a whole is defeated and death occurs.

This process and its denouement are understood in two major ways. Although they are sometimes stated as alternative formulations of a "whole brain definition" of death, they are actually mirror images of each other. The Commission has found them to be complementary; together they enrich one's understanding of the "definition." The first focuses on the integrated functioning of the body's major organ systems, while recognizing the centrality of the whole brain, since it is neither revivable nor replaceable. The other identifies the functioning of the whole brain as the hallmark of life because the brain is the regulator of the body's integration. The two conceptions are subject to similar criticisms and have similar implications for policy.

The Concepts

The functioning of many organs—such as the liver, kidney, and skin—and their integration are "vital" to individual health in the sense that if any one ceases and that function is not restored or artificially replaced, the organism as a whole cannot long survive. All elements in the system are mutually interdependent, so that the loss of any part leads to the breakdown of the whole and, eventually, to the cessation of functions in every part.

Three organs—the heart, lungs, and brain—assume special significance, however, because their interrelationship is very close and the irreversible cessation of any one very quickly stops the other two and consequently halts the integrated functioning of the organism as a whole. Because they were easily measured, circulation and respiration were traditionally the basic "vital signs." But breathing and heartbeat are not life itself. They are simply used as signs—as one window for viewing a deeper and more complex reality: a triangle of interrelated systems with the brain as its apex. As the biomedical scientists who appeared before the Commission made clear, the traditional means of diagnosing death actually detected an irreversible cessation of integrated functioning among the interdependent bodily systems. When artificial means of support mask this loss of integration as measured by the old methods, brain-oriented criteria and tests provide a new window on the same phenomena.

On this view, death is the moment at which the body's physiological system ceases to constitute an integrated whole. Even if life continues in individual cells or organs, life of the organism as a whole requires complex integration, and without the latter, a person cannot properly be regarded as alive.

This distinction between systematic, integrated functioning and physiological activity in cells or individual organs is important for two reasons. First, a person is considered dead under this concept even if oxygenation and metabolism persist in some cells or organs. There would be no need to wait until all metabolism had ceased in every part before recognizing that death has occurred.

More importantly, this concept would reduce the significance of continued respiration and heartbeat for the definition of death. This view holds that continued breathing and circulation are not in themselves tantamount to life. Since life is a matter of integrating the functioning of major organ systems, breathing and circulation are necessary but not sufficient to establish that an individual is alive. When an individual's breathing and circulation lack neurologic integration, he or she is dead.

The alternative "whole brain" explanation of death differs from the one just described primarily in the vigor of its insistence that the traditional "vital signs" of heartbeat and respiration were merely surrogate signs with no significance in themselves. On this view, the heart and lungs are not important as basic prerequisites to continued life but rather because the irreversible cessation of their functions shows the brain had ceased functioning. Other signs customarily employed by physicians in diagnosing death, such as unresponsiveness and absence of pupillary light response, are also indicative of loss of the functions of the whole brain.

This view gives the brain primacy not merely as the sponsor of consciousness (since even unconscious persons may be alive), but also as the complex organizer and regulator of bodily functions. (Indeed, the "regulatory" role of the brain in the organism can be understood in terms of thermodynamics and information theory.) Only the brain can direct the entire system. Artificial support for the heart and lungs, which is required only when the brain can no longer control them, cannot maintain the usual synchronized integration of the body. Now that other traditional indicators of cessation of brain functions (i.e., the absence of breathing) can be obscured by medical interventions, one needs, according to this view, new standards for determining death—that is, more reliable tests for the complete cessation of brain functions.

4. Not Globally Diagnosed the Same

At first glance, one might think that defining brain death using empirical methods would be a simple task in clinical medicine. One might think that all clinicians have to do is follow standard protocols in order to confirm when someone is dead. But as Dr. Eelco F. M. Wijdicks shows, defining death in clinical medicine is far from simple. Wijdicks is a member of the Department of Neurology and a practicing physician in the Neurological and Neurosurgical Intensive Care Unit of the Mayo Clinic in Rochester, Minnesota. He has spent his career investigating clinical definitions of death. Wijdicks defines *brain death* as follows:

> "Brain death" is the vernacular expression for irreversible loss of brain function. Brain death is declared when brainstem reflexes, motor responses, and respiratory drive are absent in a normothermic, nondrugged comatose patient with a known irreversible massive brain lesion and no contributing metabolic derangements. The determination of brain death in adults has become an integral part of neurologic and neurosurgical practice but may include physicians of any specialty. Institutional policies and legal provisions are in place in the United

States and elsewhere. . . . Here the results of a survey of brain death guidelines throughout the world are provided and relevant differences are considered, noting countries without formal guidelines. (2002, 20)

While many countries adopt a brain-oriented definition of death, they do not always share the same procedures for determining death. This leads Wijdicks to distinguish important differences in defining death and to call for further standardization of the clinical definition of death in medicine. What follows is the abstract to his essay that appears in *Neurology*.

Reading: "Brain Death Worldwide: Accepted Fact but No Global Consensus in Diagnostic Criteria" (Wijdicks 2002, 20). Reprinted with the permission of *Neurology*.

Abstract

Objective: To survey brain death criteria throughout the world.

Background: The clinical diagnosis of brain death allows organ donation or withdrawal of support. Declaration of brain death follows a certain set of examinations. The code of practice throughout the world has not been systematically investigated.

Methods: Brain death guidelines in adults in 80 countries were obtained through review of literature and legal standards and personal contacts with physicians.

Results: Legal standards on organ transplantation were present in 55 of 80 countries (69%). Practice guidelines for brain death for adults were present in 70 of 80 countries (88%). More than one physician was required to declare brain death in half of the practice guidelines. Countries with guidelines all specifically specified exclusion of confounders, irreversible coma, absent motor response, and absent brainstem reflexes. Apnea testing, using a PCO_2 target, was recommended in 59% of the surveyed countries. Differences were also found in time of observation and required expertise of examining physicians. Additional provisions existed when brain death was due to anoxia. Confirmatory laboratory testing was mandatory in 28 of 70 practice guidelines (40%).

Conclusion: There is uniform agreement on the neurologic examination with exception of the apnea test. However, this survey found other major differences in the procedures for diagnosing brain death in adults. Standardization should be considered.

5. Total Brain Failure

The President's Council on Bioethics was commissioned by Executive Order 13237 on November 28, 2001, by then-US president George W. Bush. Its mission is to advise the president of the United States on bioethical issues that may emerge as a consequence of advances in biomedical science and technology. Over its years of meetings, it has released publications on a variety of topics, including the ethical use of stem cells, human cloning, and new biotechnologies. It has also spent time on the problem of determining clinical death, particularly in this day and age of organ transplantation and pointed criticism that patients are not really dead when they are declared dead prior to the retrieval of organs. In what follows, the President's Council considers criticisms of the whole-brain definition of death and arrives at this conclusion: "The prevailing opinion is that the current neurological standard for declaring death, grounded in a careful diagnosis of total brain failure, is biologically and philosophically defensible" (2009, ¶2). Nevertheless, as the council says, questions remain about whether a brain-dead patient is really dead. Brain-dead patients still exhibit cellular functioning, such as hair and nail growth, wound healing, and nervous system responses. In some sense, they still are "alive."

Reading: "A Summary of the Council's Debate on the Neurological Standard for Determining Death," in *Controversies in the Determination of Death: A White House Paper by the President's Council on Bioethics* (President's Council 2009, chap. 7)

As we noted in the Preface and in Chapter One, although this report addresses several controversies in the determination of death, including those arising in the context of controlled DCD [i.e., donation after cardiac death], its primary focus is on the debates surrounding the neurological standard for the determination of death. In its deliberations, the President's Council on Bioethics did, indeed, discuss controlled DCD and the traditional cardiopulmonary standard; it also voiced concerns about the problem of ensuring adequate end-of-life care for the patient-donor. The Council's principal concern, however, was with the question, *Does a diagnosis of "whole brain death" mean that a human being is dead?* In other words, does the neurological standard rest on a sound biological and philosophical basis?

   Among members of the President's Council on Bioethics, the prevailing opinion is that the current neurological standard for declaring death, grounded in a careful diagnosis of total brain failure, is biologically and philosophically defensible. The ethical controversies explored in this report were first raised for the Council during its inquiry into organ transplantation: as most deceased organ donors have been declared dead on the basis of the neurological standard, questions about its validity have an obvious relevance for organ procurement. The Council concluded that, despite that connection, the two matters— determining death and procuring organs—should be addressed separately. More precisely, questions about the vital status of neurologically injured individuals should be taken up *prior to* and *apart from* ethical issues in organ procurement from deceased donors.

Two such questions must be posed and answered in light of certain clinical and pathophysiological facts and in light of the competing interpretations of those facts. First, *are patients in the condition of total brain failure actually dead?* And, second, *can we answer the first question with sufficient certainty to ground a course of action that treats the body in that condition as the mortal remains of a human being?* Most members of the Council have concluded that both questions can and should be answered in the affirmative. They reaffirm and support the well-established dictates of both law and practice in this area.

Many members of the Council, however, judge that affirmative answers to these questions must be supported by arguments better than and different from those offered in the past. Until now, two facts about the diagnosis of total brain failure have been taken to provide fundamental support for a declaration of death: first, that the body of a patient with this diagnosis is no longer a "somatically integrated whole," and, second, that the ability of the patient to maintain circulation will cease within a definite span of time. Both of these supposed facts have been persuasively called into question in recent years.

Another argument, however, can be advanced to support the declaration of death following a diagnosis of total brain failure. It is one that many members of the Council find both sound and persuasive, for it appeals to long recognized facts about the condition of total brain failure, while doing so in a way that is both novel and philosophically convincing. According to this argument, the patient with total brain failure is no longer able to carry out the fundamental work of a living organism. Such a patient has lost—and lost irreversibly—a fundamental openness to the surrounding environment as well as the capacity and drive to act on this environment on his or her own behalf. As described in Chapter Four, a living organism engages in self-sustaining, need-driven activities critical to and constitutive of its commerce with the surrounding world. These activities are authentic signs of active and ongoing life. When these signs are absent, and these activities have ceased, then a judgment that the organism as a whole has died can be made with confidence.

However, another view of the neurological standard was also voiced within the Council. According to this view, there can be no certainty about the vital status of patients with total brain failure; hence, the only prudent and defensible conclusion is that such patients are severely injured—but not yet dead—human beings. Therefore, only the traditional signs—irreversible cessation of heart and lung function—should be used to declare a patient dead. Also, according to this view, medical interventions for patients with total brain failure should be withdrawn only after they have been judged to be *futile,* in the sense of medically *ineffective and non-beneficial* to the patient and disproportionately *burdensome.* Such a judgment must be made on ethical grounds that consider the whole situation of the particular patient and not merely the biological facts of the patient's condition.[note i] Once such a judgment has been made, interventions can and should be withdrawn so that the natural course of the patient's injury can reach its inevitable terminus. Only after this process has occurred and the patient's heart has stopped beating, is there a morally valid warrant to proceed with such steps as preparation for burial or for organ procurement.

With this report, the President's Council on Bioethics seeks to shed light on a matter of ongoing ethical and philosophical controversy in contemporary medicine. Knowing when death has come, along with what can and should be done before and after it has arrived, has always been a problem for humankind, to one degree or another. But the nature and significance of the problem have changed over time, especially in the wake of technological advances that enable us to sustain life, or perhaps just the appearance of it, indefinitely. Given these changes and others that are yet to come, the Council believes that it is necessary and desirable to re-examine our ideas and practices concerning the human experience of death in light of new evidence and novel arguments. Undertaken in good faith, such a re-examination is a responsibility incumbent upon all who wish to keep human dignity in focus, especially in the sometimes disorienting context of contemporary medicine.

(i) This understanding of medical futility has been developed in several papers by Edmund D. Pellegrino, the Council's chairman. In these (as well as other) works, Pellegrino argues that clinical judgments of the futility of a given therapeutic intervention involve a "judicious balancing" of three factors: (1) the *effectiveness* of the given intervention, which is an objective determination that physicians alone can make; (2) the *benefit* of that intervention, which is an assessment that only patients and/or their surrogates can make; and (3) the *burdens* of the intervention (e.g., the cost, discomfort, pain, or inconvenience), which are jointly assessed by both physicians and patients and/or their surrogates. For example, see E. D. Pellegrino, "Decisions to Withdraw Life-Sustaining Treatment: A Moral Algorithm," *JAMA,* 283, no. 8 (2000): 1065–67; and E. D. Pellegrino, "Futility in Medical Decisions: The Word and the Concept," *HEC Forum,* 17, no. 4 (2005): 308–18.

C. REFLECTIONS: CLINICAL DEATH AND ORGAN DONATION

1.(*) Explore further biological death: Investigate what happens when the body dies from the standpoint of a scientist or clinician. Record what happens to the body prior to, at the moments of, and the days after death (in the case of a nonembalmed body). Summarize your findings. Be sure to give details. Share your reaction to what you have learned.

2.(*) Those who support a strictly biological account of death tend to be agnostics or atheists. Go online and find out more about the views held by agnostics and atheists. Summarize the views held by agnostics and atheists. Be sure to give details on these views and examples of proponents.

3.　As Eelco Wijdicks indicates, one of the reasons countries adopt a clinical definition of death is so that there is available an objective way to determine death for patients who are donating organs to recipients. Reflect on the decision to donate an organ. Go online and research the options to donate organs where you live (e.g., through your driver's license bureau, a regional medical center, or a museum exhibit called Body Works). Look up the specific procedures for donating organs. If you are an organ donor, why have you decided to be one? If you are not an organ donor, why have you decided not to be one? If you wish, complete the paperwork necessary to be an organ donor.

D. FURTHER READINGS

Ad Hoc Committee of the Harvard Medical School. 1968. "A Definition of Irreversible Coma: Report of the Ad Hoc Committee of the Harvard Medical School to Examine the Definition of Brain Death." *Journal of the American Medical Association* 205 (6): 337–40.

Canadian Congress Committee on Brain Death. 1988. "Death and Brain Death: A New Formulation for Canadian Medicine." *Canadian Medical Association Journal* 138:405–6.

Heron, Melonie. 2012. "Deaths: Leading Causes for 2009." *National Vital Statistics Reports* 61 (7): 1–96.

High, Dallas. 2003. "Death: Philosophical and Theological Foundations." In *Encyclopedia of Bioethics,* 3rd ed., edited by S. G. Post, vol. 1, 301–7. New York: Macmillan.

McQuoid-Mason, David. 2012. "Human Tissue and Organ Transplant Provisions: Chapter 8 of the National Act and Its Regulations, in Effect from March 2012—What Doctors Must Know." *South African Medical Journal* 102 (9): 733–35. www.samj.org.za/index.php /samj/article/view/6047.

Nuland, Sherwin. 1995. *How We Die: Reflections on Life's Final Chapter.* New York: Vintage, 1995.

Rich, Ben A. 2014. "Structuring Conversations on the Fact and Fiction of Brain Death." *American Journal of Bioethics* 14 (8): 31–33.

Wahlster, Sarah, et al. 2015. "Brain Death Declaration." *Neurology* 84:1870–79.

Wilkinson, Martin, and Stephen Wilkinson. 2016. "The Donation of Human Organs." In *Stanford Encyclopedia of Philosophy.* http://plato.stanford.edu/entries/organ-donation/.

World Health Organization. 2018. "The Top 10 Causes of Death." Fact Sheet, May. www.who.int/news-room/fact-sheets/detail/the-top-10-causes-of-death.

Youngner, Stuart, Robert M. Arnold, and Renie Shapiro, eds. 1999. *The Definition of Death.* Baltimore: Johns Hopkins University Press.

— Chapter 3 —

Psychological Disintegration

A. CONTEXT

Chapter 2 explored death as a failure of the whole brain to function. For some, a whole-brain definition of death goes "too far" in requiring a patient's entire brain to have lost integrated functioning for an individual to be declared dead. It goes too far because some individuals would consider themselves dead at the point consciousness is lost, which in this case occurs *before* the onset of whole-brain death. For others, a whole-brain definition of death does not go far enough. It does not go far enough because some individuals hold that what is unique about being human is not brain processing but the self, personal identity, or essence that defines what it means to be human and alive. Given this, the loss of brain functioning may not be sufficient for someone to be considered dead. Either way, these types of reflections lead to a different account of death, one that I call "psychological disintegration," where *psychological* means pertaining to the mind or mental processes that are not reducible to physical brain components or processes.

The previous view of death as physical disintegration differs from the view of death as psychological disintegration presented in this chapter. The view of death as physical disintegration lends itself to a physical reductionist view. A *philosophical reductionist* view holds that objects or phenomena at one level of description are explicable in terms of objects or phenomena at a more specific level of description. The view entails a number of commitments: for example, (1) one theory reduces to another (a position called *theoretical reductionism*); (2) the best strategy is to reduce descriptions and explanations to the smallest possible entities (a position called *epistemic reductionism*); and (3) the "whole" is nothing more than the sum of its parts (where the whole is not seen to be "real") (a position called *ontological reductionism*). In the case of death, a reductionist might take death to be reducible to the lack of brain functioning. The view is that (1) one theory of death (e.g., death as "psychological disintegration") reduces to another (e.g., death as "physical disintegration"); (2) the best strategy is to reduce descriptions and explanations of death (e.g., of an individual human) to the smallest possible entities (e.g., organ, tissue, or cellular disintegration); and (3) "death" is nothing more than the sum of its parts (where the parts are nonoperating tissues or cells in the brain).

In contrast, the view that death is psychological disintegration lends itself to a philosophical nonreductionist view. A *philosophical nonreductionist* challenges the view that objects or phenomena at one level of description and explanation are explicable in terms of objects or phenomena at a more specific level of description and explanation. In the case of death, death is more than a description or explanation of a bunch of nonfunctioning tissues or cells in the brain. It is the loss of consciousness, self, or personal identity—the absence of "who I am." It is the loss of that which is "essential," "unique," or "special" to humans.

A nonreductionist view of death is the type of position advanced in the readings that follow. An early version of the view that "I" am defined by my consciousness is found in John Locke's *An Essay Concerning Human Understanding*. More recently, an account of the view that the loss of one's psychological integration equates to death is found in *Defining Death*, authored by the President's Commission for the Study of Ethical Problems in Medicine and Biomedical and Behavioral Research. It is a view that is endorsed by Robert Veatch in "The Impending Collapse of the Whole-Brain Definition of Death" and by Ben Rich in "Postmodern Personhood: A Matter of Consciousness."

B. PERSPECTIVES

1. Loss of Consciousness

British philosopher and physician John Locke (1632–1704) gained recognition for his view advanced in *Second Treatise of Government* ([1689] 1980) that citizens have a right to property in the moral and civil society. This is an idea that attracted much support in political circles following its publication and continues to influence thinking in democratic societies today. Beyond developing this influential social and political notion, Locke spent time on questions concerning the nature of knowledge. Central to his view of knowledge is his account of the nature of the self as a knower.

For Locke, the self as a knower is an observer who perceives the world through empirical lenses. That is, the self observes reality through patterns of perceptions and knows reality through an interpretation of such observations. Yet, like Aristotle (see chapter 2, B.1), Locke is not a strict empiricist. He held that the self is made up of something more than sense data. Well before the popularity of philosophy of mind and accounts of personal identity, Locke held that personal identity is a matter of continued consciousness or memory. As he says, "Since consciousness always accompanies thinking, and it is that which makes every one to be what he calls self, and thereby distinguishes himself from all other thinking things, in this alone consists personal identity" (Locke 1959, 449). Although Locke does not address death per se in the passage, his view of what it means to be a human knower has implications for an understanding of death. On his view, the loss of consciousness means the loss of personal identity and, for those who equate personal identity with life, is death. In other words, death is the loss of being "a thinking intelligent being, that has reason and reflection" (448).

Reading: "Personal Identity," in *An Essay Concerning Human Understanding* (Locke 1959, 448–52). Reprinted with the permission of Dover Publications, Inc.

11. This being premised, to find wherein personal identity consists, we must consider what *person* stands for; —which, I think, is a thinking intelligent being, that has reason and reflection, and can consider itself as itself, the same thinking thing, in different times and places; which it does only by that consciousness which is inseparable from thinking, and, as it seems to me, essential to it: it being impossible for anyone to perceive without *perceiving* that he does perceive. When we see, hear, smell, taste, feel, meditate, or will anything, we know that we do so. Thus it is always as to our present sensations and perceptions: and by this every one is to himself that which he calls *self*:—it not being considered, in this case, whether the same self be continued in the same or diverse substances. For, since consciousness always accompanies thinking, and it is that which makes every one to be what he calls self, and thereby distinguishes himself from all other thinking things, in this alone consists personal identity, i.e. the sameness of a rational being: and as far as this consciousness can be extended backwards to any past action or thought, so far reaches the identity of that person; it is the same self now it was then; and it is by the same self with this present one that now reflects on it, that that action was done.

    12. But it is further inquired, whether it is the same identical substance. This few would think they had reason to doubt of, if these perceptions, with their

consciousness, always remained present in the mind, whereby the same think-ing thing would be always consciously present, and, as would be thought, evi-dently the same to itself. But that which seems to make the difficulty is this, that this consciousness being interrupted always by forgetfulness, there being no moment of our lives wherein we have the whole train of all our past actions be-fore our eyes in one view, but even the best memories losing the sight of one part whilst they are viewing another; and we sometimes, and that the greatest part of our lives, not reflecting on our past selves, being intent on our present thoughts, and in sound sleep having no thoughts at all, or at least none with that consciousness which remarks our waking thoughts, — I say, in all these cases, our consciousness being interrupted, and we losing the sight of our past selves, doubts are raised whether we are the same thinking thing, i.e. the same *sub-stance* or no. Which, however reasonable or unreasonable, does not concern *personal* identity at all. The question being what makes the same person; and not whether it be the same identical substance, which always thinks in the same person, which, in this case, matters not at all: different substances, by the same consciousness (where they do partake in it) being united into one person, as well as different bodies by the same life are united into one animal, whose iden-tity is preserved in that change of substances by the unity of one continued life. For, it being the same consciousness that makes a man be himself to himself, personal identity depends on that only, whether it be annexed solely to one indi-vidual substance, or can be continued in a succession of several substances. For as far as any intelligent being *can* repeat the idea of any past action with the same consciousness it had of it at first, and with the same consciousness it has of any present action; so far it is the same personal self. For it is by the con-sciousness it has of its present thoughts and actions, that it is *self to itself* now, and so will be the same self, as far as the same consciousness can extend to actions past or to come, and would be by distance of time, or change of sub-stance, no more two persons than a man be two men by wearing other clothes today than he did yesterday, with a long or a short sleep between: the same con-sciousness uniting those distant actions into the same person, whatever sub-stances contributed to their production.

2. Loss of Psychological Capacities and Properties

As discussed in chapter 2, B.3, the President's Commission for the Study of Ethical Problems in Medicine and Biomedical and Behavioral Research (1981) generated a well-received definition of brain death, one that has served as the basis of US state law defining death. The definition has come to be known as a "whole-brain" definition of death. In its report on death, the commission also discussed the view that a higher-brain, as opposed to a whole-brain, definition of death is preferred by some because it accounts for an important interpretation of death based on higher-brain or psychological disintegration. Note here that the commission distinguishes between higher-brain somatic integration and so-called psychological or mental integration. Although the commission did not advocate for the adoption of a higher-brain view of death, it recognized that the position has garnered support. As the President's Commission says in the following passage, higher-brain formulations of brain-oriented definitions of death "are premised on the fact that loss of cerebral functions strips the patient of his psychological capacities and properties" (1981, 38). While this interpretation views death as empirical because psychological disintegration is understood in terms of measurable correlates in the brain (i.e., loss of cerebral function), death is not entirely reducible to loss of biological function. It involves some estimate of the loss of psychological function and what defines "consciousness," "the self," or "personal identity." The President's Commission recognizes the emerging interest in "higher-brain" formulations of death but does not endorse them because it views the whole-brain definition of death as more widely supported in US medicine and culture.

Reading: *Defining Death: Medical, Legal and Ethical Issues in the Determination of Death* (President's Commission 1981, 38–39)

The "Higher Brain" Formulations [of Death]

When all brain processes cease, the patient loses two important sets of functions. One set encompasses the integrating and coordinating functions carried out principally but not exclusively by the cerebellum and brainstem. The other set includes the psychological functions which make consciousness, thought, and feeling possible. These latter functions are located primarily but not exclusively in the cerebrum, especially the neocortex. The two "higher brain" formulations of brain-oriented definitions of death discussed here are premised on the fact that loss of cerebral functions strips the patient of his psychological capacities and properties.

A patient whose brain has permanently stopped functioning will, by definition, have lost those brain functions which sponsor consciousness, feeling, and thought. Thus the higher brain rationales support classifying as dead bodies which meet "whole brain" standards, as discussed in the preceding section. The converse is not true, however. If there are parts of the brain which have no role in sponsoring consciousness, the higher brain formulation would regard their continued functioning as compatible with death.

The Concepts

Philosophers and theologians have attempted to describe the attributes a living being must have to be a person. "Personhood" consists of the complex of activities (or of capacities to engage in them) such as thinking, reasoning, [and] feeling, human intercourse which make the human different from, or superior to, animals or things. One higher brain formulation would define death as the loss of what is essential to a person. Those advocating the personhood definition often relate these characteristics to brain functioning. Without brain activity, people are incapable of these essential activities. A breathing body, the argument goes, is not in itself a person; and without functioning brains, patients are merely breathing bodies. Hence personhood ends when the brain suffers irreversible loss of function.

For other philosophers, a certain concept of "personal identity" supports a brain-oriented definition of death. According to this argument, a patient literally ceases to exist as an individual when his or her brain ceases functioning, even if the patient's body is biologically alive. Actual decapitation creates a similar situation: the body might continue to function for a short time, but it would no longer be the "same" person. The persistent identity of a person as an individual from one moment to the next is taken to be dependent on the continuation of certain mental processes which arise from brain functioning. When the brain processes cease (whether due to decapitation or to "brain death") the person's identity also lapses. The mere continuation of biological activity in the body is irrelevant to the determination of death, it is argued, because after the brain has ceased functioning the body is no longer identical with the person.

3. Irreversible Cessation of the Capacity for Consciousness

Despite widespread support for a whole-brain definition of death, there is continued interest in the need to develop a higher-brain definition of death. Robert M. Veatch is a professor of medical ethics at the Kennedy Institute of Ethics at Georgetown University in Washington, D.C., and has been writing about issues in bioethics for decades. As previously indicated, bioethics, or biomedical or health care ethics, emerged as a special field of study in the 1970s in response to the significant ethical issues raised by the development and use of biomedical interventions. As a proponent of a higher-brain definition of death, Veatch holds that "no one really believes that literally all functions of the entire brain must be lost for an individual to be dead. A better definition of death involves a higher brain orientation" (1993, 18). He notes that a higher-brain formulation of death involves measuring irreversible loss of capacity for consciousness and, as such, involves neurological criteria as well as "fundamentally nonscientific value judgments" (23) regarding consciousness and personal identity, those aspects of humanness that defy being reduced to biological functioning. The following passage elaborates on Veatch's defense of a higher-brain definition of death.

Reading: "The Impending Collapse of the Whole-Brain Definition of Death" (Veatch 1993, 18, 22–23). Reprinted with the permission of Wiley.

[p. 18]
For many years there has been lingering doubt, at least among theorists, that the currently fashionable "whole-brain-oriented" definition of death has things exactly right. I myself have long resisted the term "brain death" and will use it only in quotation marks to indicate the still common, if ambiguous, usage. The term is ambiguous because it fails to distinguish between the biological claim that the brain is dead and the social/legal/moral claim that the individual as a whole is dead because the brain is dead. An even greater problem with the term arises from the lingering doubt that individuals with dead brains are really dead. Hence, even physicians are sometimes heard to say that the patient "suffered brain death" one day and "died" the following day. It is better to say that he "died" on the first day, the day the brain was determined to be dead, and that the cadaver's other bodily functions ceased the following day. For these reasons I insist on speaking of persons with dead brains as individuals who are dead, not merely persons who are "brain dead."

The presently accepted standard definition, the Uniform Determination of Death Act, specifies that an individual is dead who has sustained "irreversible cessation of all functions of the entire brain, including the brain stem." It also provides an alternative definition specifying that an individual is also dead who has sustained "irreversible cessation of circulatory and respiratory functions." The President's Commission for the Study of Ethical Problems in Medicine and Biomedical and Behavioral Research made clear, however, that circulatory and respiratory function loss are important only as indirect indicators that the brain has been permanently destroyed.
[pp. 22–23]

Crafting New Public Law

. . . Two changes would be needed in the current definition of death: (1) incorporating the higher brain function notion and (2) incorporating some form of the conscience clause.

Present law makes persons dead when they have lost all functions of the entire brain. It is uniformly agreed that the law should incorporate only this basic concept of death, not the precise criteria or tests needed to determine that the whole brain is dead. That is left up to the consensus of neurological experts.

All that would be needed to shift to a higher brain formulation is a change in the wording of the law to replace "all functions of the entire brain" with some relevant, more limited alternative. There are at least three options: references to higher brain functions, cerebral functions, or consciousness. While we could simply change the wording to read that an individual is dead when there is irreversible cessation of all higher brain functions, that poses a serious problem. We are now suffering from the problems created by the vagueness of the definition referring to "all functions of the entire brain." Even though referring to "all higher brain functions" would be conceptually correct, it would be even more ambiguous. It would lack needed specificity.

This specificity could be achieved by referring to irreversible loss of cerebral functions, but we have already suggested two problems with that wording. Just as we now know there are some isolated functions of the whole brain that should be discounted, so there are probably some isolated cerebral functions that most would not want to count either. For example, if, hypothetically, an isolated "nest" of cerebral motor neurons were perfused so that if stimulated the body could twitch, that would be a cerebral function, but not a significant one for determining life any more than a brain stem reflex is. Second, in theory some really significant functions such as consciousness might some day be maintainable even without a cerebrum—if, for example, a computer could function as an artificial center for consciousness. [*Ed. note: see further discussion of this idea in chapter 7 of this volume.*] The term "cerebral function" adds specificity but is not satisfactory.

The language that seems best if integration of mind and body is what is critical is "irreversible cessation of the capacity for consciousness." That is, after all, what the defenders of the higher brain formulations really have in mind. (If someone were to claim that some other "higher" function is critical, that alternative could simply be plugged in.) As is the case now, the specifics of the criteria and tests for measuring irreversible loss of capacity for consciousness would be left up to the consensus of neurological expertise, even though measuring irreversible loss of capacity for a brain function such as consciousness involves fundamentally nonscientific value judgments. If the community of neurological expertise claims that irreversible loss of consciousness cannot be measured, so be it. We will at least have clarified the concept and set the stage for the day when it can be measured with sufficient accuracy. We have noted, however, that neurologists presently claim they can in fact measure irreversible loss of consciousness accurately.

4. Loss of Conscious Experience

Like Veatch (see previous section in this chapter), Ben A. Rich supports a higher-brain definition of death. Rich is an attorney with a doctoral degree in philosophy. He currently teaches in the Bioethics Program at the University of California at Davis and oversees the training of medical staff, students, and other professionals. Rich has substantial experience both in private practice and as a legal counsel for several leading academic medical centers. His special interests include the legal and ethical aspects of the physician-patient relationship, end-of-life care, and pain management. In what follows, Rich argues that the President's Commission "fails to provide a meaningful basis upon which to distinguish between the death of a human being and any other species of animal" (1997, 211). He offers this: "What is unique about human beings is their capacity for personhood, for living the self-conscious life of a person. It should be, therefore, the total and permanent loss of that capacity that marks the death of the human being" (212). On this view, death is the loss of the capacity for conscious experience and the end of human personal life or consciousness.

Reading: "Postmodern Personhood: A Matter of Consciousness" (Rich 1997, 210–12). Reprinted with permission of Wiley.

At this point, we must return to the President's Commission discussion of brain death. Through its insistence upon a whole brain formulation of death, i.e., cessation of function of the entire brain, including the brain stem, the Commission rejects high order consciousness or any consciousness at all, for that matter, as the essence of personhood. The brain is treated not as the organ which sponsors the rational, continuing self-conscious life of the person, but rather as the organ which engenders the body's capacity to organize and regulate itself. Of primary concern here are the homeostasis of body temperature, heartbeat, blood pressure, respiration, and the like. Thus, according to the Commission, human death is "that moment at which the body's physiological system ceases to constitute a regulated whole" [President's Commission 1981, 33]. Through such an analysis, the capacity for conscious experience is reduced to a non-essential, indeed a trivial aspect of human life, and hence is allowed to play no role in the determination of death.

However, there has developed in bioethics an important distinction between human biological life and human personal life. Such a distinction is unintelligible outside of the realm of beings with the capacity not merely for consciousness, but for self-consciousness. Moreover, those who place great emphasis upon that distinction, such as James Rachels and H. Tristram Engelhardt, carefully note that these two dimensions of the lives of human beings are not co-extensive.

There are, they maintain, many cases in the medical domain in which human personal life comes to an end before, in some instances long before, human biological life. There are also cases in which human biological life comes into existence, but without the capacity to sponsor the human personal life that is characterized by consciousness. The President's Commission appears to dispute this claim, or at the very least to attach no significance to it whatsoever. Yet a patient in a persistent vegetative state, who has undergone what is referred to as high brain, cerebral, or neocortical death and whose life as a person has ended, may, through the continued application of heroic measures, have their bodily functions sustained for many years.

I take the position that the Commission's whole brain formulation of death fails to provide a meaningful basis upon which to distinguish between the death of a human being and any other species of animal. All such animals have the capacity to auto-regulate their physiological systems as an integrated whole, but none but human beings have ever been considered to have the capacity for personhood. While the Commission might have taken the position that although we do not know how persons are distinguishable from other animals, our capacity at this time to accurately determine the permanent loss of the capacity for self-conscious experience is insufficient, that is not what it did. Instead, it dismissed the relevance of the concept of person to human death, and proceeded to recommend a formulation of death (the whole brain) that could just as easily be applied to any animal whose brain is sufficiently developed to enable it to organize and regulate its overall physiological functioning.

The whole brain death formulation has led to a great deal of ambiguity and anxiety in end of life decisionmaking. Patients who have permanently lost the capacity for conscious experience, but who have a functioning brain stem, cannot be treated as brain dead, but must be allowed to die in appropriate cases through the removal of life support. Too often, even patients who do not meet the whole brain death criterion are described by medical personnel and lay persons alike as have been declared brain dead at time t, when life support was removed, and as having "died" thereafter at time t1. This conceptual confusion is, I suggest, a product of the reduction of persons to the integrated functioning of the human body that is inherent in the whole brain formulation of death. The infirmity of the whole brain formulation becomes readily apparent when the Commission attempts to articulate what it is that makes a patient in a permanent vegetative state, but with a functioning brain stem, a living human being while a patient who has undergone whole brain death but is being sustained through artificial nutrition and hydration and mechanical ventilation is nothing more than a "perfused corpse."

The critical distinction, according to the Commission, the vital signs of human life present in the former patient but absent in the latter, are that the former can "breathe (without the aid of a respirator), sigh, yawn, track light with their eyes, and react to painful stimulation." . . .

Robert Veatch, in my judgment, is correct when he suggests that death should mean "a complete change in the status of a living entity characterized by the irretrievable loss of those characteristics that are essentially significant to it" [1976, 25]. Such an approach to the conceptualization of death, it would seem, is based upon the premise that the unique attributes of the living organism should carry through and inform the determination that it has died. As I earlier suggested, what is unique about human beings is not their capacity to auto-regulate their physiology, a trait which they share with many other non-human species. What is unique about human beings is their capacity for personhood, for living the self-conscious life of a person. It should be, therefore, the total and permanent loss of that capacity that marks the death of the human being.

C. REFLECTIONS: CLINICAL DEATH AND THE COST OF A FUNERAL OR MEMORIAL SERVICE

1.(*) Part of the challenge of a higher-brain definition of death is that it relies on concepts, such as consciousness, that can be difficult to define. Lend a hand to advocates of a higher-brain definition of death and help them define consciousness. What is consciousness? What does it mean to be self-conscious? Be sure to develop the view that you advance.

2.(*) In light of your readings in this chapter, take a position: Do you support a whole-brain definition of death (discussed in chapter 2, B.3 and B.5) or a higher-brain definition of death (discussed in this chapter, B.2) or some other account? Be sure to develop your reasoning.

3. This reflection returns you to your own funeral or memorial plans outlined in chapter 1. Now that you have made your choices in chapter 1, D.3, about what kind of funeral or memorial service you would like, it is time to think about the financial costs.

    The National Funeral Directors Association in the United States estimates that the "median cost of a funeral with viewing and burial for calendar year 2014 is $7,181. If a vault is included, something that is typically required by a cemetery, the median cost is $8,508. The cost does not take into account cemetery, monument, or marker costs or miscellaneous cash-advance charges, such as for flowers or an obituary" (National Funeral Directors 2016). This cost represents a 28.6 percent increase in funeral costs from the prior decade.

    Your task is to come up with a written estimate of the total cost of your funeral or memorial service. Use the service that you have already designed in your reflections found in chapter 1, D.3. Perhaps this will help you get started. The cost of the funeral or memorial service is typically made up of the following:

    a. body preparation
    b. burial vault
    c. burial shroud
    d. casket
    e. clergy/minister/official
    f. cemetery costs
    g. death certificates (need ten or so)
    h. direct burial at a cemetery
    i. direct cremation
    j. embalming
    k. flowers
    l. monument or marker costs
    m. music
    n. funeral ceremony fees
    o. funeral home services
    p. grave marker
    q. grave opening and closing

r. grave plot

s. gravesite service

t. hearse

u. obituary in local, regional, national, or international news outlet

v. printed materials

w. procession

x. urn

y. vault

z. visitation or viewing

In your reflections, list your expenses and include a total. What do your expenses say about what is important to you when you die?

D. FURTHER READINGS

Carter, Chris. 2010. *Science and the Near-Death Experience: How Consciousness Survives Death.* Rochester, VT: Inner Traditions.

Chalmers, David. 2002. *Philosophy of Mind: Classical and Contemporary Readings.* Oxford: Oxford University Press.

DeGrazia, David. 1998. "Biology, Consciousness, and the Definition of Death." *Report from the Institute of Philosophy and Public Policy* 18 (1–2): 18–22. https://www.researchgate.net/publication/11695025_Biology_consciousness_and_the_definition_of_death.

Engelhardt, H. Tristram, Jr. 1996. *The Foundations of Bioethics.* 2nd ed. New York: Oxford University Press.

Lizza, John. 2006. *Persons, Humanity, and the Definition of Death.* Baltimore: Johns Hopkins University Press.

Lommel, Pim van. 2010. *Consciousness beyond Life: The Science of the Near-Death Experience.* New York: HarperOne.

Moody, Raymond. 2001. *Life after Life: The Investigation of a Phenomenon—Survival of Bodily Death.* New York: HarperOne.

Nuland, Sherwin. 1995. *How We Die: Reflections on Life's Final Chapter.* New York: Vintage.

Rachels, James. 1996. *The Ends of Life: Euthanasia and Morality.* Oxford: Oxford University Press.

Veatch, Robert M. 1976. *Death, Dying, and the Biological Revolution.* New Haven, CT: Yale University Press.

Youngner, Stuart, Robert M. Arnold, and Renie Shapiro, eds. 1999. *The Definition of Death.* Baltimore: Johns Hopkins University Press.

— Chapter 4 —

Reincarnation

A. CONTEXT

Reincarnation, also known as metempsychosis or transmigration of the soul, re-fers to a state in which the soul or spirit survives death and is reborn in a new form or embodiment (Barry 2007, 111). Reincarnation is a widely held belief in the world, especially in Asia. A classic form of reincarnation appeared in India in the Hindu Brāhmana writings around the ninth century BCE. The later Hindu writings in the Upanishads and the Bhagavad Gita defined the concept of reincarnation in greater detail. Advocates of other Asian spiritualities that originated in India (e.g., by Buddhists and Jainists), China (e.g., by Chinese Taoists), and Tibet (e.g., by Tibetan Buddhists) adopted and developed the concept in various ways ("Reincarnation" 2017).

Belief in reincarnation is also found in the West. The Greek philosopher Soc-rates, as reported by his student Plato in the *Republic*, suggested that the soul preexists in the celestial world, "lives" in the human body for a duration of time, and, if purified, returns to a state of pure being. Found here is an early formulation of reincarnation not typically associated with Platonic thought. While reincarnation is not a mainstream belief held widely in the West, there is evidence of support for it in the traditions of the Zuñis of North America and the Senegalese of West Africa, as well as in the writings of the early Greek philosopher and mathematician Pytha-goras and the Neoplatonist Plotinus (Barry 2007, 115).

Among the reincarnationists, there are various expressions of beliefs and views about death. For example, some Hindus and Tibetan Buddhists hold that reincarnation involves a personal transmigration of the soul. Others, such as Buddhists, Taoists, and some Hindus, reject the view that there is a personal metaphysical essence or soul, yet advance the view that death involves continued, but not personal, existence or oneness with the divine (Barry 2007, 116). On this view, death involves an impersonal merging with the divine, leading some thinkers to prefer the term *rebirth* or *mystical union* to *reincarnation*. Further, there are discussions among those who believe in reincarnation about when and how reincarnation occurs. For instance, Tibetan lamas or spiritual leaders hold that reincarnation occurs immediately after death, while other reincarnationists believe that reincarnation can take centuries. Some reincarnationists hold that humans may be reborn as nonhuman animals, while others hold that humans cannot reanimate as nonhuman animals (Barry 2007, 111). Still further, there are notable differences among reincarnationists with regard to death rituals. Hindu, Tibetan Buddhist, Japanese Buddhist, and Confucian rituals are notably different with regard to their practices and prayers. Once again, there is no single story about death.

The readings that follow offer a small sample of views on reincarnation. Readers are encouraged to explore further the conceptual and practical details of the reincarnation traditions found here as well as in other traditions not represented here. Readers are encouraged to keep in mind that no tradition is simple and that there are complex and varying ways reincarnationists view death.

B. PERSPECTIVES

1. Asian Views

a. Everlasting Peace

Our exploration of reincarnation begins with an account found in Hindu thought. Hinduism is the religion of the majority of people in India and Nepal. In Hinduism, the Upanishads (800–300 BCE) consist of commentary on the Vedas. The Vedas (ca. 1500–600 BCE) represent the oldest and most authoritative ancient sacred writings of Hinduism. The word *Veda* comes from the Sanskrit word meaning "knowledge." The Vedas are organized into four collections, namely, the *Rigveda*, *Yajurveda*, *Samaveda*, and *Atharvaveda*. These consist of sacred hymns and incantations and teach a doctrine of order and reciprocity. The Upanishads are also known as the Vedanta, the end of the Veda. The Upanishads contain truth about ultimate reality (*brahman*) and describe the character and form of human salvation (*moksha*). A primary message of the Upanishads is that *atman* (i.e., the Ultimate Self) is *brahman* (i.e., Absolute Reality, Pure Being, or Consciousness). The Upanishads have more than two hundred parts composed between the early centuries BCE through the modern period. As with all the spiritual traditions, there are notable differences in interpretation and application of sacred texts.

The *Katha Upanishad* is believed to have been composed after the fifth century BCE and explores the question of what remains at death. The text contains the teachings of Yama, the Lord of Death, concerning true immortality. It retells the story of the boy Nachiketa, who seeks knowledge about life after death and learns that "the embodied soul whose dwelling is the body dissolves and from the body is released" (*Katha Upanishad* 1996, 179). For those who achieve *atman*, the following is reported to occur:

> Permanent among impermanents, conscious among the conscious.
> The One among the many, Disposers of desires:
> Wise men who see Him as subsistent in [their] selves [or self-subsistent]
> Taste of everlasting peace—no others.
>
> (*Katha Upanishad* 1996, 180)

For those who do not achieve *atman*, the following is reported to occur:

> Some to the womb return—
> Embodied souls, to receive another body;
> Others pass into a lifeless stone (*sthāṇu*)
> In accordance with their works (*karma*),
> In accordance with [the tradition] they had heard (*śruta*).
>
> (*Katha Upanishad* 1996, 179)

The following selection from the *Katha Upanishad* shares what happens at death. The immortality promised here is an impersonal merging with or absorption into the divine. "This in truth is That" sums it up: "That" is *brahman* or "Pure Being," and "this" is *atman* or the "Ultimate Self." In other words, *atman* is *brahman*; ultimate self is pure being.

V.

1. Whoso draws nigh to the city of the eleven gates [i.e., the body]
Of him who is not born, whose thought is not perverse,
He grieves not, for he has won deliverance:
Deliverance is his!

This [*atman* or Ultimate Self] in truth
is That [*brahman* or Pure Being]!

2. As swan he dwells in the pure [sky],
As god (*vasu*) he dwells in the atmosphere,
As priest he dwells by the altar,
As guest he dwells in the house:
Among men he dwells, in vows,
In Law (*rta*) and in the firmament;
Of water born, of kine, of Law (*rta*),
Of rock—[He], the great cosmic Law (*rta*)!

3. He leads the out-breath upward
And casts the in-breath downward:
To this Dwarf seated at the centre
All gods pay reverence.

4. When the embodied soul whose dwelling is the body
Dissolves and from the body is released,
What then of this remains?

This in truth is That.

5. Neither by breathing in nor yet by breathing out
Lives any mortal man:
By something else they live
On which the two [breaths] depend.

6. Lo! I will declare to thee this mystery
Of Brahman never-failing,
And of what the self becomes
When it comes to [the hour of] death.

7. Some to the womb return—
Embodied souls, to receive another body;
Others pass into a lifeless stone (*sthāṇu*)

In accordance with their works (*karma*),
In accordance with [the tradition] they had heard (*śruta*).

8. When all things sleep, [that] Person is awake,
Assessing all desires:
That is the Pure, that Brahman,
That the Immortal, so they say:
In It all the worlds are stablished;
Beyond it none can pass.

This in truth is That.

9. As the one fire ensconced within the house
Take on the forms of all that's in it,
So the One Inmost Self of every being
Takes on several forms, [remaining] without [the while].

10. As the one wind, once entered into a house,
Takes on the forms of all that's in it,
So the One Inmost Self of every being
Takes on their several forms, [remaining] without [the while].

11. Just as the sun, the eye of all the world,
Is not defiled by the eye's outward blemishes,
So the one Inmost Self of every being
Is not defiled by the suffering of the world—
[But remains] outside [it].

12. One and all mastering is the Inmost Self of every being;
He makes the one form manifold;
Wise men who see Him as subsistent in [their] selves,
Taste everlasting joy (*sukha*)—no others.

13. Permanent among impermanents, conscious among the conscious.
The One among the many, Disposers of desires:
Wise men who see Him as subsistent in [their] selves
Taste of everlasting peace—no others.

14. "That is this," so think [the wise]
Concerning that all-highest bliss which none can indicate.

How, then, should I discern it?
Does It shine of itself or but reflect the brilliance?

15. There the sun shines not, nor moon nor stars;
These lightenings shine not [there]—let alone this fire.
All things shine with the shining of this light,
This whole world reflects its radiance.

b. Never to Be Born Again

Another sacred text of Hinduism is the Bhagavad Gita. The Bhagavad Gita is part of the Mahabharata, which contains the great history of the descendants of Bharata. It dates back to the last half of the first century BCE, around 400 BCE (Goodall 1996, xxi–xxvi). The Gita, as it is often called, is compared with Homer's *Iliad* and *Odyssey* in Greek thought and the Bible in Christian thought in its influence on thought through the ages and its conveying of major philosophical and moral messages. It is a seven-hundred-stanza poem in the form of a dialogue between Lord Krishna and his disciple, the Pandava prince Arjuna. The poem covers a range of fundamental topics concerning existence and the meaning of life.

In the poem, Arjuna is about to go into battle with his own cousins. The message revealed in the dialogue is that God (Vishnu) has incarnated himself in the human form of Krishna to teach divine truth. Krishna advises Arjuna about the duties of a warrior and elaborates on numerous topics, including death, salvation, and reincarnation. Arjuna asks Krishna several questions about the path to salvation. Here is Krishna's reply:

15. Coming right nigh to Me, these great of soul,
Are never born again.
For rebirth is full of suffering, knows nothing that abides:
[Free from it now] they attain the all-highest prize (*saṃsiddhi*).

16. The worlds right up to Brahmā's realm
[Dissolve and] evolve again;
But he who comes right nigh to Me
Shall never be born again.

(*Bhagavad-Gita* 1996, 244 [§8])

What follows is a selection from section 8 of the Gita featuring Krishna's account of what happens at death and how "he who comes to Me" is never born again. Note that the immortality promised in the Gita is personal.

Reading: "Shall Never Be Born Again," in the Bhagavad Gita (*Bhagavad-Gita* 1996, 243–45 [§8]). © 1996 by J. M. Dent. Berkeley: University of California Press. Reprinted with permission of University of California Press.

[Krishna speaking]
5. Whoso at the hour of death,
Abandoning his mortal fame,
Bears Me in mind and passes on,
Accedes to my Divinity (*mad-bhāva*): have no doubt of that.

6. Whatever state (*bhāva*) a man may bear in mind
When the time comes at last to cast the mortal frame aside,
Even to that state does he accede,
For ever does that state of being make him grow into itself
(*tadbhāva-bhāvita*).

7. Then muse upon Me always,
And go to war;
For if thou fixest mind and soul (*buddhi*) on Men,
To Me shalt thou most surely come.

8. Let a man's thoughts be integrated with the discipline (*yoga*)
Of constant striving: let them not stray to anything else [at all];
So by meditating on the divine All-Highest Person,
[That man to that All Highest] goes.

9. For [He it is who is called] the Ancient Seer,
Governor [of all things, yet] smaller than the small,
Ordainer of all, in form unthinkable,
Sun-Coloured beyond the darkness. Let a man meditate on Him [as such].

10. With mind unmoving when his turn comes to die,
Steadied (*yukta*) by loyal love (*bhakti*) and Yogic power,
Forcing the breath between the eyebrows duly,
[So will such a man] draw nigh to the divine All-Highest Person.

11. The imperishable state [word] of which the Vedic scholars speak,
Which sages enter, [all their] passion spent,
For love of which men lead a life of chastity,
[That state] will I proclaim, to thee in brief.

12. Let a man close up all [the body's] gates,
Stem his mind within his heart,
Fix his breath within his head,
Engrossed in Yogic concentration.

13. Let him utter [the word] Oṁ, Brahman in one syllable,
Keeping me in mind;
Then when his time is come to leave aside the body,
He'll tread the highest Way.

14. How easily am I won by him
Who bears Me in mind unceasingly,
Thinking of nothing else at all—
A Yogin integrated ever.

15. Coming right nigh to Me, these great of soul,
Are never born again.
For rebirth is full of suffering, knows nothing that abides:
[Free from it now] they attain the all-highest prize [saṁsiddhi].

16. The worlds right up to Brahmā's realm
[Dissolve and] evolve again;
But he who comes right nigh to Me
Shall never be born again.

17. For a thousand ages lasts
One day of Brahmā,
And for a thousand ages one such night:
This knowing, men will know [what is meant by] day and night.

18. At the day's dawning all things manifest
Spring forth from the Unmanifest;
And then at nightfall they dissolve again
In [that same mystery] surnamed "Unmanifest."

19. Yea, this whole host of beings
Comes ever anew to be; at fall of night
It dissolves away all helpless;
At dawn of day it rises up [again].

20. But beyond that there is [yet] another mode of being—
Beyond the Unmanifest another Unmanifest, eternal:
This is He who passes not away
When all contingent beings pass away.

21. Unmanifest surnamed "Imperishable"—
This, men say, is the All-Highest Way,
And this once won, there is no more returning:
This is my all-highest home [*dhāma*].

c. Liberation and Enlightenment

Buddhism is a tradition dating back to the sixth century BCE and the teachings of the Indian Siddhartha (ca. 560–480 BCE), or Gautama after the name of the family. Guatama lived in Kapilavastu in the Indian subcontinent, which is present-day Nepal. According to tradition, Gautama was not simply human. He was a human destined to greatness, a "Buddha" or "Awakened One" (Bilhartz 2006, 207). The Buddha left no writings, but his teachings, such as those found in the *Dhammapada*, were recorded by his followers. Buddha challenged orthodox Indian teaching and rejected the view that there is a pure metaphysical essence, such as a soul or *atman*. Rather, for Buddha, there is void, or no-self. About five centuries after the life of Gautama, Buddhism entered a second era of creative development with the rise of Mahayana Buddhism and an emphasis on achieving *bodhisattva* (i.e., an enlightened [*bodhi*] existence [*sattva*]). Five centuries after this marks the rise of Tantric Buddhist literature, including *The Tibetan Book of the Dead* (*Bardo Thodol Chenmo*), one of the great treatises on the nature of death and how to die (Bilhartz 2006, 210–14).

According to tradition, *The Tibetan Book of the Dead* was written by eighth-century spiritual master Padmasambhava and was discovered in the twelfth century CE in Tibet by Rigzin Karma Linpa, a spiritual master believed to be his reincarnation. The book is a self-help guidebook intended to help the dying and recently deceased souls to find their way through the difficult stages of the afterlife. In the text, "bardos" are existential gaps between stages that must be addressed. The first part, called *Chikhai Bardo*, describes the moment of death. The second part, called *Chonyid Bardo*, addresses the states that supervene immediately after death. The third part, called *Sidpa Bardo*, concerns the onset of the birth instinct and of prenatal events. The gaps or uncertainties in and after life must be eliminated so that the dying person can journey on without the illusion of attachment. The cycle of re-births is the natural result of various good deeds and misdeeds in the person's life. It can involve innumerable lives, including a journey in either sex, in nonhuman animals, and in other realms. It inevitably involves suffering and continues until all cravings are lost. Then and only then, as the following passage indicates, "Having recognized himself, he will become inseparably united with the dharmakāya [Perfect Enlightenment] and certainly attain liberation" (*Tibetan Book of the Dead* 1990, 87). On one account, the following selection from *The Tibetan Book of the Dead* is intended to be read to the dying person by a wise person (e.g., a guru, dharma-brother, or dharma-sister) who guides the dying person in the journey to the afterlife. Instructions to the wise person appear below, and passages read to the dying person appear in quotation marks.

Reading: "The Great Liberation through Hearing in the Bardo," in *The Tibetan Book of the Dead* (1992, 81–87). © 1975 by Francesca Fremantle and Chögyam Trungpa. Reprinted by arrangement with The Permissions Company, Inc., on be-half of Shambhala Publications, Inc., Boulder, Colorado, www.shambhala.com.

The time of instruction: when respiration has ceased, prāṇa [i.e., breath or breathing] is absorbed into the wisdom-dhūtī and luminosity free from complexities shines clearly in the consciousness. If prana is reversed and escapes into the right and left nāḍīs [i.e., energy channel], the bardo state appears suddenly, so the reading should take place before the prāṇa escapes into the right and left nāḍīs. The length of time during which the inner pulsation remains after respiration has ceased is just about the time to take to eat a meal.

The method of instruction: it is best if ejection of consciousness is effected when the respiration is just about to stop, but if it has not been affected one should say these words:

"O child of noble family, (name), now the time has come for you to seek a path. As soon as your breath stops, what is called the basic luminosity of the first bardo, which your guru has already shown you, will appear to you. This is the dharmatā [i.e., intrinsic nature], open and empty like space, luminous voice, pure naked mind without center or circumference. Recognize then, and rest in that state, and I too will show you at the same time."

This should be firmly implanted in his mind by repeating it many times over in his ear until he stops breathing. Then, when the ceasing of the breath is heard, one should lay him down on the right side in the lion position and firmly press the two pulsating arteries, which induce sleep, until they stop throbbing. Then the prāṇa which has entered the dhūtī will not be able to go back and will be certain to emerge through the brahmarandhara.

Now the showing should be read. At this time the first bardo, which is called the luminosity of dharmatā, the undistorted mind of the dharmakāya [i.e., Perfect Enlightenment] arises in the mind of all beings. Ordinary people call this state unconscious because the prāṇa sinks into the avadhūtī [i.e., energy channels] during the interval between the ceasing of the breath and the pulsation. The time it lasts is uncertain, depending on the spiritual condition and the stage of yogic training. It lasts for a long time in those who have practiced much, were steady in the meditation practice of tranquility, and sensitive. In striving to show such a person one should repeat the instruction until pus comes out from the apertures of his body. In wicked and insensitive people it does not last longer than a single snapping of the fingers, but in some it lasts for the time taken to eat a meal. As most sūtras [i.e., rules or aphorisms] and tantras [i.e., doctrine of enlightenment] say that this unconscious state lasts for four and a half days, generally one should strive to show the luminosity for that length of time.

The method of instruction: If he is able, he will work with himself from the instructions already given. But if he cannot by himself, then his guru, or a disciple of his guru, or a dharma-brother or dharma-sister who was a close friend, should stay nearby and read aloud clearly the sequence of the signs of death: "Now the sign of earth dissolving into water is present, water into fire, fire into air, air into consciousness. . . ." When the sequence is almost complete he should be encouraged to adopt an attitude like this, "O child of noble family," or, if he was a guru, "O Sir,"—"do not let your thoughts wander." This should be spoken softly in his ear. In the case of a dharma-brother, a dharma-sister, or anyone else, one should call him by name and say these words:

"O child of noble family, that which is called death has now arrived, so you should adopt this attitude: 'I have arrived at the time of death, so now, by means of this death, I will adopt only the attitude of the enlightened state of mind, friendliness and compassion, and attain perfect enlightenment for the sake of all sentient beings as limitless as space. With this attitude, at this special time for the sake of all sentient beings, I will recognize the luminosity of death as the dharmakāya, and attaining in that state the supreme realization of the Great Symbol, I will act for the good of all beings. If I do not attain this, I will recognize the bardo state as it is, and attaining the indivisible Great Symbol form in the bardo, I will act for the good of all beings as limitless as space in whatever way will influence them.' Without letting go of this attitude you should remember and practice whatever meditation teaching you have received in the past."

These words should be spoken distinctly with the lips close to his ear, so as to remind him of his practice without letting his attention wander even for a moment. Then, when respiration has completely stopped, one should firmly press the arteries of sleep and remind him with these words, if he was a guru or spiritual friend higher than oneself:

"Sir, now the basic luminosity is shining before you; recognize it, and rest in the practice."

And one should show all others like this:

"O child of noble family, (name), listen. Now the pure luminosity of the dharmatā is shining before you; recognize it. O child of noble family, at this moment your state of mind is by nature pure emptiness, it does not possess any nature whatever, neither substance nor quality such as color, but it is pure emptiness; this is the dharmatā, the female Buddha Samantabhadrī. But this state of mind is not just blank emptiness, it is unobstructed, sparkling, pure and vibrant; this mind is the male Buddha Samantabhadra. These two, your mind whose nature is emptiness without any substance whatever, and your mind which is vibrant and luminous, are inseparable: this is the dharmakāya of the Buddha. This mind of yours is inseparable luminosity and emptiness in the form of a great mass of light, it has no birth or death, therefore it is the Buddha of Immortal Light. To recognize this is all that is necessary. When you recognize this pure nature of your mind as the Buddha, looking into your own mind is resting in the Buddha-mind."

This should be repeated three or seven times, clearly and precisely. Firstly, it will remind him of what he has previously been shown by his guru; secondly, he will recognize his own naked mind as the luminosity; and thirdly, having recognized himself, he will become inseparably united with the dharmakāya and certainly attain liberation.

2. Western Views

a. Separation of the Soul from the Body

Students in philosophy are often surprised that the Western ancient Greek philosopher Plato (ca. 427–348 BCE) supported reincarnationist views of death. Plato was the student of Socrates (ca. 470–399 BCE) and teacher of Aristotle (384–322 BCE) (see chapter 2, B.1, for more on Aristotle). Plato is known for his systematic analysis of fundamental questions about reality, knowledge, and the ethical and political life. His *Republic* is recognized as one of the great works in Western literature. In it, Plato addresses the nature of reality, how we know it, and what is required to live the good life.

In the *Republic*, in a selection that is referred to as "the Myth of Er," Plato recounts the story of soldier Er, who was killed on the battlefield, was allowed to see the afterlife, and awoke from the funeral pile to tell his story. Er reports his obser-vations about judgment, reward, and punishment and advances an early version of reincarnation. As stated in the *Republic*, "He said that when his soul left the body it went on a journey with a great company, and that they came to a mysterious place at which there were two openings in the earth; they were near together, and over against them were two other openings in heaven above" (Plato 1953b, 491).

In the intermediate space between earth and heaven, there were judges seated. The judges were figures "who commanded the just, after they had given judgement on them and had bound their sentences in front of them, to ascend by the way up through heaven on the right hand; and in like manner the unjust were bidden by them to descend by the lower way on the left hand; these also bore tokens of all their deeds, but fastened on their backs" (492). According to Plato, death is marked by the separation of the soul from the body, the soul is judged according to how one leads one's life, and the "unjust" souls "descend" to a new existence on earth. Although the notion of reincarnation found here was not widely adopted in the West, the view that death is marked by the soul separating from the body and is accompanied by some type of judgment gained significant support in Abrahamic (i.e., Jewish, Chris-tian, and Islamic) traditions of thought (see chapter 5 for more on this).

Reading: *Republic* (Plato 1953b, 491–93 [bk. 10]). Reprinted with the permis-sion of Oxford Publishing Limited.

Well, I said, I will tell you a tale; not one of the tales which Odysseus tells to the hero Alcinous, yet this too is a tale of a hero, Er the son of Armenius, a Pam-phylian by birth. He was slain in battle, and ten days afterwards, when the bod-ies of the dead were taken up already in a state of corruption, his body was found unaffected by decay, and carried away home to be buried. And on the twelfth day, as he was lying on the funeral pile, he returned to life and told them what he had seen in the other world. He said that when his soul left the body it went on a journey with a great company, and that they came to a mysterious place at which there were two openings in the earth; they were near together, and over against them were two other openings in heaven above. In the interme-diate space there were judges seated, who commanded the just, after they had given judgement on them and had bound their sentences in front of them, to as-cend by the way up through heaven on the right hand; and in like manner the un-just were bidden by them to descend by the lower way on the left hand; these also bore tokens of all their deeds, but fastened on their backs. He drew near, and they told him that he was to be the messenger who would carry the report of the other world to men, and they bade him hear and see all that was to be heard and seen in that place. Then he beheld and saw on one side the souls departing at either opening of heaven and earth when sentence had been given on them; and at the two other openings other souls, some ascending out of the earth dusty and worn with travel, some descending out of heaven clean and bright. And arriving ever and anon they seemed to have come from a long journey, and they went forth with gladness into the meadow, where they encamped as at a festival; and those who knew one another embraced and conversed, the souls which came from earth curiously inquiring about the things above, and the souls which came from heaven about the things beneath. And they told one another of what had happened by the way, those from below weeping and sorrowing at the remembrance of the things which they had endured and seen in their journey beneath the earth (now the journey lasted a thousand years), while those from above were describing heavenly delights and visions of inconceivable beauty. The full story, Glaucon, would take too long to tell; but the sum was this:—He said that for every wrong they had done and every person whom they had injured they had suffered tenfold; or once in a hundred years—such being reckoned to be the length of a man's life, and the penalty being thus paid ten times in a thousand years. If, for example, there were any who had been the cause of many deaths by the betrayal of cities or armies, or had cast many into slavery, or been accessory to any other ill treatment, for all their offenses, and on behalf of each man wronged, they were afflicted with tenfold pain, and the rewards of beneficence and justice and holiness were in the same proportion. I need hardly repeat what he said concerning young children dying almost as soon as they were born. Of piety and impiety to gods and parents, and of murder, there were retributions other and greater from which he described.

b. A Sacred Place

The Zuñi is a federally recognized Native American tribe in the United States. Most of the members of the Zuñi tribe live in the Pueblo of Zuñi on the Zuñi River, a tributary of the Little Colorado River, in western New Mexico in the United States. In addition, the tribe owns trust lands in Catron County, New Mexico, and Apache County, Arizona. Spirituality is central to Zuñi life. Zuñi spiritual belief focuses on the three most powerful of their deities, namely, Earth Mother, Sun Father, and Moonlight-Giving Mother, as well as other *kachinas* or *kokos*, which means "rain spirits" ("Zuni," 2017).

According to tradition, all Zuñi ancestors are reborn as clouds and rain (Chidester 2002, 70). Because ancestors are reincarnated in these different forms, the Zuñis offer food to the dead in ceremonies honoring the dead and pray to the clouds and rain for life and well-being. As a Zuñi prayer says,

> This day my children,
> For their fathers,
> Their ancestors,
> For you who have attained the far off place of waters [i.e., the dead],
> This day
> My children
> Have prepared food for your rite.
>
> (Bunzel 1929–30, 621–22)

This spiritual and practical way of remaining in communion with the dead acknowledges that the dead are part of the spiritual essence of the universe. Here, the dead are not just one's own but all those in the tribal ancestry. Because the living remain in communion with the dead, it is important to be on good terms with them with the hope that the deceased will grant the descendants good things. As the prayer says,

> Therefore I have added to your hearts.
> To the end, my fathers,
> My children:
> You will protect us.
> All my ladder descending children
> Will finish their roads;
> They will grow old.
> You will bless us with life.
>
> (Bunzel 1929–30, 623)

The following prayer is an offering of food to the ancestors and conveys how the Zuñi people view death as a sacred place within the universe.

Reading: "An Offering of Food to the Ancestors," in *Zuñi Ritual Poetry* (Bunzel 1929–30, 621–23)

[p. 621]
This day my children,
For their fathers,
Their ancestors,
[p. 622]
5 For you who have attained the far off place of waters,[note 3]
This day
My children
Have prepared food for your rite.
10 Now our sun father
Has gone in to sit down at his sacred place.[note 4]
Taking the food my children have prepared at their fireplaces
(I have come out.)
15 Those who hold our roads,[note 5]
The night priests,[note 6]
Coming out rising to their sacred place
Will pass us on our roads.
20 This night
I add to your hearts.
With your supernatural wisdom
You will add to your hearts,
Let none of you be missing
25 But all add to your hearts.
Thus on all sides you will talk together.
From where you stay quietly
Your little wind-blown clouds,
Your fine wisps of cloud,
30 Your massed clouds you will send
forth to sit down with us;
With your fine rain caressing the earth,
With all your waters
You will pass to us on our roads.
With your great pile of waters,
35 With your fine rain caressing the earth,
With your heavy rain caressing the earth,
You will pass to us on our roads.
My fathers,
Add to your hearts.

40 Your waters,
Your seeds,
Your long life,[note 7]
Your old age
You will grant to us.
[p. 623]
45 Therefore I have added to your hearts.
To the end, my fathers,
My children:
You will protect us.
All my ladder descending children[note 8]
50 Will finish their roads;
They will grow old.
You will bless us with life.

------

3. That is, the dead.
4. The sun has two resting places: one above, to which he "comes out standing" at sunrise; one below the world, to which he "goes in to sit down" at sunset.
5. Used for any supernaturals who influence human affairs. This is not a special deity. . . .
6. That is the night itself, anthropomorphically envisaged.
7. Literally "road fulfilling."
8. That is, human, the inhabitants of Zuñi.

c. Not Being Gone

Senegalese poet and storyteller Birago Diop (1906–89) was a renowned veterinarian and diplomat known for his folktales and poetic expressions of Native African beliefs. He served as an ambassador to Tunisia from the western African nation of Senegal between 1961 and 1965. One of Diop's famous poems retells a widely held African belief that the dead are reincarnated, ever present, and found in different forms here on earth. The poem, which was originally published in 1960 in *Leurres et Lueurs* (2000), with an English translation in 1961 (Jahn 1961), emphasizes the unbroken biological and spiritual continuity of ancestors who continue to be with the living (Chidester 2002, 53). As the poem says, "Those who are dead are never gone. . . . The dead are not dead" (Diop 1991, 232). On this view, ancestors or the "living dead" constitute a vital part of the identity of the community.

Reading: "The Dead Are Never Gone" (Diop 1991, 232–33)

Those who are dead are never gone;
They are there in the thickening shadow.
The dead are not under the earth:
They are there in the tree that rustles,
They are in the wood that groans,
They are in the water that runs,
They are in the water that sleeps,
They are in the hut, they are in the crowd,
The dead are not dead.

Those who are dead are never gone.
They are in the breasts of the women,
They are in the child who is wailing,
and in the firebrand that flames.
The dead are not under the earth;
They are in the fire that is dying,
They are in the grasses that weep,
They are in the whimpering rocks,
They are in the forest, they are in the house,
The dead are not dead.

C. REFLECTIONS: ASIAN FUNERAL OR MEMORIAL SERVICE

1. Observe and analyze an Asian funeral or memorial service: Choose a particular Asian tradition, go online, and view its funeral or memorial service. (You will be able to find one by googling, for instance, "Buddhist funeral service video.") Explain what function the ritual serves in the tradition. Select at least *three* aspects of the service that relate to death and explain the meaning of each within the particular Asian tradition. Be sure to give details in your analysis.

2.(*) Reflect further on how reincarnation is understood in the readings and in any additional reincarnationist account you may wish to draw upon. In your own words, what is reincarnation? How would you explain reincarnation to someone who was not familiar with the concept? Be sure to develop your thoughts on the matter and give details.

3.(*) Do you believe that reincarnation occurs? Why or why not? Be sure to develop your thoughts on the matter and give details.

D. FURTHER READINGS

Barry, Vincent. 2007. *Philosophical Thinking about Death and Dying*. Belmont, CA: Thomson Wadsworth.

Bilhartz, Terry D. 2006. *Sacred Words: A Source Book on the Great Religions of the World*. New York: McGraw Hill.

Chidester, David. 2002. *Patterns of Transcendence: Religion, Death, and Dying*. Belmont, CA: Wadsworth.

Edwards, Paul. 2001. *Reincarnation: A Critical Examination*. New York: Prometheus Books.

Goswami, Amit. 2001. *Physics of the Soul: The Quantum Book of Living, Dying, Reincarnation, and Immortality*. Charlottesville, VA: Hampton Roads.

Jahn, Janheinz. 1961. *Muntu: An Outline of the New African Culture*. Translated by Marjorie Grene. New York: Grove Press.

"Reincarnation." 2011. In *Encyclopedia of Death and Dying*. www.deathreference.com/Py -Se/Reincarnation.html.

Smith, Huston. 1991. *The World's Religions*. New York: HarperSanFrancisco.

Stevenson, Ian. 1997. *Where Reincarnation and Biology Intersect*. New York: Praeger.

Valea, Ernest. 2012. "Reincarnation: Its Meaning and Consequences." *Many Paths to One Goal? A Comparative Analysis of the Major World Religions from a Christian Perspective*, June 16. www.comparativereligion.com/reincarnation.html.

— Chapter 5 —

Resurrection

A. CONTEXT

Resurrection refers to a state in which a living being comes back to life after death. It involves a metaphysical change in the body prior to its reunification with the soul. Accounts of resurrection are found in the Abrahamic traditions of Judaism, Christianity, and Islam. These monotheistic traditions share the view that there is a single Creator who has a personal relationship with those who are created. Upon death, the soul *and* body survive and are reborn in keeping with the miracles that are made possible through divine intervention. Judaism, Christianity, and Islam are considered "Abrahamic" traditions because all trace their origins to the patriarch Abraham. For Jews, Abraham was the first to enter into a covenant with the God of ancient Israel, who is called Yahweh (Genesis 17:7). Abraham is considered ancestor of the Israelites through his son Isaac, who was born to Sarah. Christians see Abraham as a role model of faith to God and some see Abraham as a direct ancestor of Jesus (Romans 4:9–12). Muslims believe that the prophet Muhammad is a descendent of Abraham through Abraham's son Ishmael, who was born to Hagar (Qur'an 4:163).

Although there are robust similarities among Jewish, Christian, and Muslim views of resurrection, there are differences between and within the traditions. For instance, Orthodox Jews believe that bodily resurrection will occur in the Messianic Age. When the Messiah comes, the souls of the righteous will be resurrected. Christians understand resurrection in terms of the ascension of the incarnated (i.e., God-made-human) Jesus Christ into heaven. Christ's rise from the dead indicates what is to come for the believer after death. Muslims believe that the Day of Resurrection (*Yawn al-Qiyāmah*) comes to those who follow Allah's laws. At a designated time unknown to the Muslim believer, Allah recreates the decayed body (Surah 17:100) of those who engage in good deeds.

There are also differences within particular Abrahamic traditions. Orthodox, Reform, and Conservative Jews have different understandings of death and different expectations of the funeral service. Orthodox Christians understand the funeral to be one of their sacraments. Eastern Orthodox Christians believe that the dead person is absolved of any sins for which he or she has repented. Many Roman Catholics believe in the existence of purgatory and the role prayers play in the salvation of persons in purgatory. Sunni, Shia, and Sufi Muslims differ on matters of interpretation and what constitutes proper ritual in matters of death. In this way, death takes on different meanings within particular traditions.

While resurrection is a common theme in the Abrahamic religions, it is found in other traditions as well. Accounts of resurrection appear in the Middle Eastern traditions of Egypt, Mesopotamia, and ancient Persia and in the stories of the Egyptian Lord of the Dead Osiris, the immortal Utnapishtim, and the prophet Zarathustra, respectively. All of these traditions develop early versions of resurrection in the history of ideas.

The following selections highlight a small sample of views about resurrection from Middle Eastern and Abrahamic traditions. Readers are encouraged to explore further the conceptual and practical details of the traditions found here as well as traditions not represented here. Readers are encouraged to keep in mind that no tradition is simple and that resurrectionists view death in complex and varying ways. Once again, there is no single story of death.

B. PERSPECTIVES

1. Middle Eastern Myths

a. Being Everlastingly

Like *The Tibetan Book of the Dead* (see chapter 4, B.1.c), *The Egyptian Book of the Dead* was designed to assist the deceased in his or her journey to the next world. *The Egyptian Book of the Dead* refers to a collection of papyrus texts written on scrolls during the New Kingdom period of Egypt between 1580 and 1090 BCE (Chidester 2002, 149). *The Papyrus of Nu*, which was acquired by the Trustees of the British Museum in the year 1888, is the largest, best preserved, and best illuminated of all the papyri. It was made for a scribe named Nu, who held an important religious office in the royal priesthood of Thebes, Greece, around 1300 BCE (Chidester 2002, 150). The book details the process of judgment that a person could expect in the great hall of Ma'āt, a place that upheld the divine standard of ethical judgment of the gods. In the hall, the gods determined whether a person could enter the throne room of Osiris, the Egyptian Lord of the Dead, as a glorified spirit (*akh*) in the court of Osiris. Early Egyptians believed that Osiris was resurrected after he died, and his resurrection provided the hope that others would be resurrected. A successful resurrection depended on meeting the ethical standard of Ma'āt, which was objective, impersonal, and absolute. An individual either survived as a godlike being in the halls of Osiris or was annihilated by Am-mut, the eater of the dead (Chidester 2002, 151).

What follows is from chapter 154 of the *Egyptian Book of the Dead*'s *Papyrus of Nu*. It is a testimony about what happens to an individual who successfully passes through the judgment of the dead and whose life is restored after death through resurrection. As the passage says, "Do thou embalm these my members, for I would not perish and come to an end [but would be] even like unto my divine father Khepera, the divine type of him that never saw corruption [bodily disintegration]" (2008, 518). The passage expresses the hope that Nu will not decay, rot, putrefy, turn into worms, or see bodily corruption, but rather will live and flourish in the land of "everlastingness" (2008, 518).

Reading: "The Chapter of Not Letting the Body Perish," from *The Papyrus of Nu* in *The Egyptian Book of the Dead* (2008, 518–20 [chap. 154])

The Osiris Nu saith:— Homage to thee, O my divine father Osiris! I come to embalm thee. Do thou embalm these my members, for I would not perish and come to an end [but would be] even like unto my divine father Khepera, the divine type of him that never saw corruption. Come then, strengthen my breath, O Lord of the winds, who dost magnify these divine beings who are like unto thyself. Stablish me, stablish me, and fashion me strongly, O Lord of the funeral chest. Grant thou that I may enter into the land of everlastingness, according to that which was done for thee, along with thy father Tem, whose body never saw corruption, and who himself never saw corruption. I have never done that which thou hatest, nay, I have acclaimed thee among those who love thy KA [the spiritual aspect of the soul]. Let not my body become worms, but deliver thou me as thou didst deliver thyself. I pray thee, let me not fall into rottenness, as thou lettest every god, and every goddess, and every animal, and every reptile, see corruption, when the soul hath gone out of them, after their death.

And when the soul hath departed, a man seeth corruption, and the bones of his body crumble away and become stinking things, and the members decay one after the other, the bones crumble into a helpless mass, and the flesh tur- neth into foetid liquid. Thus man becometh a brother unto the decay which cometh upon him, and he turneth into a myriad of worms, and he becometh nothing but worms, and an end is made of him, and perisheth in the sight of the god of day (Shu), even as do every god, and every goddess, and every bird, and every fish, and every creeping worm, and every reptile, and every beast, and every thing whatsoever. Let [all the Spirits fall] on their bellies [when] they recog- nize me, and behold, the fear of me shall terrify them; and thus also let it be with every being that hath died, whether it be animal, or bird, or fish, or worm, or rep- tile. Let life [rise out of] death. Let not the decay caused by any reptile make an end [of me], and let not [enemies] come against me in their various forms. Give thou me not over to the Slaughterer in this execution-chamber, who killeth the members, and maketh them rot, being [himself] invisible, and who destroyeth the bodies of the dead, and liveth by carnage. Let me live, and perform his order; I will do what is commanded by him. Give me not over to his fingers, let him not overcome me, for I am under thy command, O Lord of the Gods.

Homage to thee, O my divine father Osiris, thou livest with thy members. Thou didst not decay, thou didst not become worms, thou didst not wither, thou didst not rot, thou didst not putrefy, thou didst not turn into worms. I am the god Khepera, and my members shall have being everlastingly. I shall not decay, I shall not rot, I shall not putrefy, I shall not turn into worms, and I shall not see cor- ruption before the eye of the god Shu. I shall have my being, I shall have my being; I shall live, I shall live; I shall flourish, I shall flourish, I shall flourish, I shall wake up in peace, I shall not putrefy, my intestines shall not perish, I shall not suffer injury. My eye shall not decay. The form of my face shall not disappear. My ear shall not become deaf. My head shall not be separated from my neck. My tongue shall not be removed. My hair shall not be cut off. My eyebrows shall not be shaved away, and no evil defect shall assail me. My body shall be stablished. It shall neither become a ruin, nor be destroyed on this earth.

b. A Secret of the Gods

The Mesopotamian story *The Epic of Gilgamesh* dates back to around 2000 BCE (Chidester 2002, 152). Mesopotamia is an ancient region of Southwest Asia in present-day Iraq between the Tigris River and the Euphrates River. The tablets on which *Gilgamesh* was written were excavated in the mid-nineteenth century. *Gilgamesh* wrestles with the unavoidable fact of human mortality and explores the possibilities of resurrection of the dead. It features the story of Gilgamesh, the renowned king of Uruk, who goes into battle with his dear friend and fellow soldier Enkidu. As the story goes, Enkidu dies, and Gilgamesh grieves greatly for his current loss as well as his own future loss of life as a human. Human death in the Mesopotamian tradition is grim and unreturned by any hope of salvation, regardless of the individual's personal contributions in life. Seeking something better, Gilgamesh decides to petition the help of Utnapishtim, who, along with his wife, is the only human to achieve "everlasting life" through resurrection. While he ultimately does not attain immortality because of his own choice to return to his kingdom of Uruk and die as its king, Gilgamesh learns that immortality may be possible for certain humans and thereby gives hope to others. What follows is a retelling of Gilgamesh's encounter with Utnapishtim and what he learns about achieving immortality through resurrection. According to the story, the gods become angry with humans and wish to exterminate them. Because Utnapishtim is in good favor with the gods, one of them warns him of impending doom for humans. The god Ea shares with Utnapishtim the "secret of the gods," the way to "permanence" or "everlasting life": "Tear down your house and build a boat, abandon possessions and look for life, despise worldly goods and save your soul alive. Tear down your house, I say, and build a boat. These are the measurements of the barque [type of sailing vessel] as you shall build her: let her beam equal her length, let her deck be roofed like the vault that covers the abyss; then take up into the boat the seed of all living creatures" (*Epic of Gilgamesh* 1972, 108). According to legend, Utnapishtim follows Ea's advice and, as a consequence, achieves immortality through resurrection.

Reading: "The Search for Everlasting Life" and "The Story of the Flood," in *The Epic of Gilgamesh* (1972, 97, 105–8 [bks. 4 and 5]). Reproduced by permission of Penguin Books Ltd.

[Bk. 4] The Search for Everlasting Life

Bitterly Gilgamesh wept for his friend Enkidu; he wandered over the wilderness as a hunter, he roamed over the plains; in his bitterness he cried, "How can I rest, how can I be at peace? Despair is in my heart. What my brother is now, that shall I be when I am dead. Because I am afraid of death I will go as best I can to find Utnapishtim whom they call the Faraway, for he has entered the assembly of the gods." So Gilgamesh travelled over the wilderness, he wandered over the grasslands, a long journey, in search of Utnapishtim, whom the gods took after the deluge; and they set him to live in the land of Dilmun, in the garden of the sun; and to him alone of men they gave everlasting life. . . .

[after a long journey]

So Urshanabi [Utnapishtim's ferryman] the ferryman brought Gilgamesh to Utnapishtim. . . .

Now Utnapishtim, where he lay at ease, looked into the distance and he said in his heart, musing to himself, "Why does the boat sail here without tackle and mast; why are the sacred stones destroyed, and why does the master not sail the boat? That man who comes is none of mine; where I look I see a man whose body is covered with skins of beasts. Who is this who walks up the shore behind Urshanabi, for surely he is no man of mine?" So Utnapishtim looked at him and said, "What is your name, you who come here wearing the skins of beasts, with your cheeks starved and your face drawn? Where are you hurrying to now? For what reason have you made this great journey, crossing the seas whose passage is difficult? Tell me the reason for your coming."

He replied, "Gilgamesh is my name. I am from Uruk, from the house of Anu." Then Utnapishtim said to him, "If you are Gilgamesh, why are your cheeks so starved and your face drawn? Why is despair in your heart and your face like the face of one who has made a long journey? Yes, why is your face burned with heat and cold; and why do you come here, wandering over the wilderness in search of the wind?"

Gilgamesh said to him, "Why should not my cheeks be starved and my face drawn? Despair is in my heart and my face is the face of one who has made a long journey. It was burned with heat and with cold. Why should I not wander over the pastures? My friend, my younger brother who seized and killed the Bull of Heaven and overthrew Humbaba in the cedar forest, my friend who was very dear to me and endured dangers beside me, Enkidu, my brother whom I loved, the end of mortality has overtaken him. I wept for him seven days and nights till the worm fastened on him. Because of my brother I am afraid of death; because of my brother I stray through the wilderness. His fate lies heavy upon me. How can I be silent, how can I rest? He is dust and I shall die also and be laid in the earth for ever." Again Gilgamesh said, speaking to Utnapishtim, "It is to see Uf-napishtim whom we call the Faraway that I have come this journey. For this I have wandered over the world, I have crossed many difficult ranges, I have crossed the seas, I have wearied myself with travelling; my joints are aching, and I have lost acquaintance with sleep which is sweet. My clothes were worn out before I came to the house of Siduri. I have killed the bear and hyena, the lion and panther, the tiger, the stag and the ibex, all sorts of wild game and the small creatures of the pastures. I ate their flesh and I wore their skins; and that was how I came to the gate of the young woman [Siduri], the maker of wine, who barred her gate of pitch and bitumen against me. But from her I had news of the journey; so then I came to Urshanabi the ferryman, and with him I crossed over the waters of death. Oh, father Utnapishtim, you who have entered the assembly of gods, I wish to question you concerning the living and the dead, how shall I find the life for which I am searching?"

Utnapishtim said, "There is no permanence. Do we build a house to stand for ever, do we seal a contract to hold for all time? Do brothers divide an inheritance to keep for ever, does the flood-time of rivers endure? It is only the nymph of the dragon-fly who sheds her larva and sees the sun in his glory. From the days of old there is no permanence. The sleeping and the dead, how alike they are, they are like a painted death. What is there between the master and the servant when both have fulfilled their doom? When the Anunnaki, the judges, come together, and Mammetun the mother of destinies, together they decree the fates of men. Life and death they allot but the day of death they do not disclose."

Then Gilgamesh said to Utnapishtim the Faraway, "I look to you now, Utnapishtim, and your appearance is no different from mine; there is nothing strange in your features. I thought I should find you like a hero prepared for battle, but you lie here taking your ease on your back. Tell me truly, how was it that you came to enter the company of gods and to possess everlasting life?" Utnapishtim said to Gilgamesh, "I will reveal to you a mystery, I will tell you a secret of the gods."

[Bk. 5] The Story of the Flood

[Utnapishtim] "You know the city of Shurrupak, it stands on the banks of Euphrates? That city grew old and the gods that were in it were old. There was Anu, lord of the firmament, their father, and warrior Enlil their counsellor, Ninurta the helper, and Ennugi watcher over canals; and with them also was Ea. In those days the world teemed, the people multiplied, the world bellowed like a wild bull, and the great god was aroused by the clamour. Enlil heard the clamour and he said to the gods in council, "The uproar of mankind is intolerable and sleep is no longer possible by reason of the babel." So the gods agreed to exterminate mankind. Enlil did this, but Ea because of his oath warned me in a dream. He whispered their words to my house of reeds, "Reed-house, reed-house! Wall, O wall, hearken reed-house, wall reflect; O man of Shurrupak, son of Ubara-Tutu; tear down your house and build a boat, abandon possessions and look for life, despise worldly goods and save your soul alive. Tear down your house, I say, and build a boat. These are the measurements of the barque [type of sailing vessel] as you shall build her: let her beam equal her length, let her deck be roofed like the vault that covers the abyss; then take up into the boat the seed of all living creatures."

c. Continued Life

Zarathustra (or Zoroaster) lived between 1400 and 1200 BCE (Chidester 2002, 154) in the eastern part of ancient Iran, then called Persia, and founded the religion and philosophy called Zoroastrianism. Zarathustra preached the principle of order, truth, and righteousness manifested in good thoughts, words, and deeds. This principle is known as *asha*, which is upheld by the good or illuminating god, Ahura Mazda, and challenged by the evil god or the Lie, Angra Mainyu. The sacred scripture of Zoroastrianism is the *Avesta*, which dates back to the eighth century BCE and is composed of six parts: *Yasna, Visperad, Yashts, Vendidad*, minor texts, and fragments.

The *Yasna* part of the *Avesta* speaks about life after death and the reward of resurrection for those who are judged to be good. As *Yasna* 30:7 says, "And to him (i.e.[,] mankind) came Dominion, and Good Mind, and Right and Piety gave continued life to their bodies and indestructibility, so that by thy retributions through (molten) metal he may gain the prize over the others" (*Avesta* 1995). The judgment is automatic and personal; that is, it takes place immediately and depends on the thoughts, words, and deeds of the individual. What follows is a selection from the *Yasna* on the possibility of achieving everlasting life through resurrection.

Reading: *Avesta: Yasna: Sacred Liturgy and Gathas/Hymns of Zarathushtra* (*Avesta* 1995, *Yasna* 30). Reprinted with the permission of Soonoo Taraporewaia.

1. Now I will proclaim to those who will hear the things that the understanding man should remember, for hymns unto Ahura and prayers to Good Thought; also the felicity that is with the heavenly lights, which through Right shall be beheld by him who wisely thinks.

2. Hear with your ears the best things; look upon them with clear-seeing thought, for decision between the two Beliefs, each man for himself before the Great consummation, bethinking you that it be accomplished to our pleasure.

3. Now the two primal Spirits, who reveal themselves in vision as Twins, are the Better and the Bad, in thought and word and action. And between these two the wise ones chose aright, the foolish not so.

4. And when these twain Spirits came together in the beginning, they created Life and Not-Life, and that at the last Worst Existence shall be to the followers of the Lie, but the Best Existence to him that follows Right.

5. Of these twain Spirits he that followed the Lie chose doing the worst things; the holiest Spirit chose Right, he that clothes him with the massy heavens as a garment. So likewise they that are fain to please Ahura Mazda by dutiful actions.

6. Between these twain the Daevas also chose not aright, for infatuation came upon them as they took counsel together, so that they chose the Worst Thought. Then they rushed together to Violence, that they might enfeeble the world of men.

7. And to him (i.e. mankind) came Dominion, and Good Mind, and Right and Piety gave continued life to their bodies and indestructibility, so that by thy retributions through (molten) metal he may gain the prize over the others.

8. So when there cometh their punishment for their sins, then, O Mazda, at Thy command shall Good Thought establish the Dominion in the Consummation, for those who deliver the Lie, O Ahura, into the hands of Right.

9. So may we be those that make this world advance, O Mazda and ye other Ahuras, come hither, vouchsafing (to us) admission into your company and Asha, in order that (our) thought may gather together while reason is still shaky.

10. Then truly on the (world of) Lie shall come the destruction of delight; but they who get themselves good name shall be partakers in the promised reward in the fair abode of Good Thought, of Mazda, and of Right.

11. If, O ye mortals, ye mark those commandments which Mazda hath ordained—of happiness and pain, the long punishment for the follower of the Druj, and blessings for the followers of the Right—then hereafter shall it be well.

2. Abrahamic Accounts

a. To Be Awakened to Everlasting Life

Resurrectionist views are found pervasively in Abrahamic traditions of thought, namely, Judaism, Christianity, and Islam. Believers in Judaism subscribe to the view that there is a single, personal, and moral God, called Yahweh, who reveals himself in nature and history. The sacred texts of Judaism date back three millennia. The Torah includes the five Books of Moses: Genesis, Exodus, Leviticus, Numbers, and Deuteronomy. Some Jews also consider Nevi'im and Ketubim sacred texts. Nevi'im includes eight books that are divided in terms of the Early and Later Prophets, such as Samuel and Ezekiel, respectively. Ketubim includes eleven books that are referred to as the Writings (e.g., Daniel) (Bilhartz 2006, 19–21). Some call the Torah plus Nevi'im and Ketubim the Hebrew Bible.

Passages in Ezekiel and Daniel provide two examples of resurrection in the Jewish tradition. Ezekiel is believed to have been written in the late sixth century BCE. As Ezekiel 37:5 says, "Thus saith the Lord GOD unto these bones: Behold, I will cause breath to enter into you, and ye shall live" (*Hebrew Bible* 2002, "Ezekiel Chapter 37"). Here resurrection comes about by the "breath" of God, where "breath" is understood as the holy spirit that gives life its essence. Daniel is believed to have been written in the mid-sixth century BCE. As Daniel 12:2 says, "And many of them that sleep in the dust of the earth shall awake, some to everlasting life, and some to reproaches and everlasting abhorrence" (*Hebrew Bible* 2002, "Daniel Chapter 12"). Here resurrection involves everlasting life with the divine, assuming that one's life has been judged to be good. The following passages from Ezekiel and Daniel provide a glimpse into early Jewish views of resurrection as well as early accounts of views about resurrection.

Readings: Books of Ezekiel and Daniel, in *The Hebrew Bible in English* (2002, Ezek. 37:1–14 and Dan. 12:1–3). © 2002 all rights reserved to Mechon Mamre for this HTML version Hebrew Bible, translated by the Jewish Publication Society.

[Ezek. 37:1–14]
1 The hand of the LORD was upon me, and the LORD carried me out in a spirit, and set me down in the midst of the valley, and it was full of bones; 2 and He caused me to pass by them round about, and, behold, there were very many in the open valley; and, lo, they were very dry. 3 And He said unto me: "Son of man, can these bones live?" And I answered: "O Lord GOD, Thou knowest." 4 Then He said unto me: "Prophesy over these bones, and say unto them: O ye dry bones, hear the word of the LORD: 5 Thus saith the Lord GOD unto these bones: Behold, I will cause breath to enter into you, and ye shall live. 6 And I will lay sinews upon you, and will bring up flesh upon you, and cover you with skin, and put breath in you, and ye shall live; and ye shall know that I am the LORD." 7 So I prophesied as I was commanded; and as I prophesied, there was a noise, and behold a commotion, and the bones came together, bone to its bone. 8 And

I beheld, and, lo, there were sinews upon them, and flesh came up, and skin covered them above; but there was no breath in them. 9 Then said He unto me: "Prophesy unto the breath, prophesy, son of man, and say to the breath: Thus saith the Lord GOD: Come from the four winds, O breath, and breathe upon these slain, that they may live." 10 So I prophesied as He commanded me, and the breath came into them, and they lived, and stood up upon their feet, an exceeding great host. 11 Then He said unto me: "Son of man, these bones are the whole house of Israel; behold, they say: Our bones are dried up, and our hope is lost; we are clean cut off. 12 Therefore prophesy, and say unto them: Thus saith the Lord GOD: Behold, I will open your graves, and cause you to come up out of your graves, O My people; and I will bring you into the land of Israel. 13 And ye shall know that I am the LORD, when I have opened your graves, and caused you to come up out of your graves, O My people. 14 And I will put My spirit in you, and ye shall live, and I will place you in your own land; and ye shall know that I the LORD have spoken, and performed it, saith the LORD."

[Dan. 12:1–13]
1 And at that time shall Michael stand up, the great prince who standeth for the children of thy people; and there shall be a time of trouble, such as never was since there was a nation even to that same time; and at that time thy people shall be delivered, every one that shall be found written in the book. 2 And many of them that sleep in the dust of the earth shall awake, some to everlasting life, and some to reproaches and everlasting abhorrence. 3 And they that are wise shall shine as the brightness of the firmament; and they that turn the many to righteousness as the stars for ever and ever.

4 But thou, O Daniel, shut up the words, and seal the book, even to the time of the end; many shall run to and fro, and knowledge shall be increased. 5 Then I Daniel looked, and, behold, there stood other two, the one on the bank of the river on this side, and the other on the bank of the river on that side. 6 And one said to the man clothed in linen, who was above the waters of the river: "How long shall it be to the end of the wonders?" 7 And I heard the man clothed in linen, who was above the waters of the river, when he lifted up his right hand and his left hand unto heaven, and swore by Him that liveth for ever that it shall be for a time, times, and a half; and when they have made an end of breaking in pieces the power of the holy people, all these things shall be finished. 8 And I heard, but I understood not; then said I: "O my Lord, what shall be the latter end of these things?"

9 And he said: "Go thy way, Daniel; for the words are shut up and sealed till the time of the end. 10 Many shall purify themselves, and make themselves

white, and be refined; but the wicked shall do wickedly; and none of the wicked shall understand; but they that are wise shall understand. 11 And from the time that the continual burnt-offering shall be taken away, and the detestable thing that causes appalment set up, there shall be a thousand two hundred and ninety days. 12 Happy is he that waiteth, and cometh to the thousand three hundred and five and thirty days. 13 But go thou thy way till the end be; and thou shalt rest, and shalt stand up to thy lot, at the end of the days."

b. To Be Raised from the Dead

Along with Jews, Christians believe in resurrection. Christianity began in the first century. Believers in Christianity subscribe to the view that there is a single, personal, and moral God who reveals himself directly. More specifically, God reveals himself through his Son Jesus Christ, who was born on earth and later crucified and resurrected from the dead in order to redeem sinful humanity from its fallen state. The sacred texts of Christianity are books in the Bible, which includes the Old Testament (or, as some call the books, the Hebrew Bible) and the New Testament. The Old Testament includes the twenty-four books of Jewish scripture, which Christians subdivide differently, as well as other books, which vary among Christian denominations. The New Testament includes the twenty-seven books written in the century following the death of Jesus. Chief among these are the four Gospels (i.e., Matthew, Mark, Luke, and John), the Epistles (e.g., Corinthians), and the Apocalypse (i.e., the Book of Revelation) (Bilhartz 2006, 63–65).

In the Christian tradition, Christ's triumph over sin and death is shared by those who have faith in the risen Christ (Bilhartz 2006, 63). As the angel says to the women after Christ was crucified, "Do not be afraid; I know that you are looking for Jesus who was crucified. 6 He is not here; for he has been raised, as he said. Come, see the place where he lay. 7 Then go quickly and tell his disciples, 'He has been raised from the dead, and indeed he is going ahead of you to Galilee; there you will see him.' This is my message to you." (Matthew 28:5–7, *New Revised Standard Version Bible* 1989). The following passages from Matthew recount Christ's death and resurrection, which serve as a central focus for Christians in their understanding of afterlife.

Readings: Book of Matthew (*New Revised Standard Version Bible* 1989, Matt. 27:45–66 and 28:1–20). (Note: The New Revised Standard Version Bible is considered a Protestant translation of the Bible. There are a number of denominations in Christianity, most notably Orthodox, Roman Catholic, and Protestant. Both Protestants and Roman Catholics consider the Bible a sacred text; they differ on matters concerning interpretation, religious authority, customs, and holy days.)

[The Death of Jesus: Matt. 27:45–66]

45 From noon on, darkness came over the whole land until three in the afternoon. 46 And about three o'clock Jesus cried with a loud voice, "Eli, Eli, lema sabachthani?" that is, "My God, my God, why have you forsaken me?" 47 When some of the bystanders heard it, they said, "This man is calling for Elijah." 48 At once one of them ran and got a sponge, filled it with sour wine, put it on a stick, and gave it to him to drink. 49 But the others said, "Wait, let us see whether Elijah will come to save him."

50 Then Jesus cried again with a loud voice and breathed his last. 51 At that moment the curtain of the temple was torn in two, from top to bottom. The earth shook, and the rocks were split. 52 The tombs also were opened, and many bodies of the saints who had fallen asleep were raised. 53 After his resurrection they came out of the tombs and entered the holy city and appeared to many. 54 Now when the centurion and those with him, who were keeping watch over Jesus, saw the earthquake and what took place, they were terrified and said, "Truly this man was God's Son!" 55 Many women were also there, looking on from a distance; they had followed Jesus from Galilee and had provided for him. 56 Among them were Mary Magdalene, and Mary the mother of James and Joseph, and the mother of the sons of Zebedee.

57 When it was evening, there came a rich man from Arimathea, named Joseph, who was also a disciple of Jesus. 58 He went to Pilate and asked for the body of Jesus; then Pilate ordered it to be given to him. 59 So Joseph took the body and wrapped it in a clean linen cloth 60 and laid it in his own new tomb, which he had hewn in the rock. He then rolled a great stone to the door of the tomb and went away. 61 Mary Magdalene and the other Mary were there, sitting opposite the tomb. 62 The next day, that is, after the day of Preparation, the chief priests and the Pharisees gathered before Pilate 63 and said, "Sir, we remember what that impostor said while he was still alive, 'After three days I will rise again.' 64 Therefore command the tomb to be made secure until the third day; otherwise his disciples may go and steal him away, and tell the people, 'He has been raised from the dead,' and the last deception would be worse than the first." 65 Pilate said to them, "You have a guard of soldiers; go, make it as secure as you can." 66 So they went with the guard and made the tomb secure by sealing the stone.

[The Resurrection of Jesus: Matt. 28:1–20]

1 After the sabbath, as the first day of the week was dawning, Mary Magdalene and the other Mary went to see the tomb. 2 And suddenly there was a great earthquake; for an angel of the Lord, descending from heaven, came and rolled back the stone and sat on it. 3 His appearance was like lightning, and his clothing white as snow. 4 For fear of him the guards shook and became like dead men. 5 But the angel said to the women, "Do not be afraid; I know that you are looking for Jesus who was crucified. 6 He is not here; for he has been raised, as he said. Come, see the place where he lay. 7 Then go quickly and tell his disciples, 'He has been raised from the dead, and indeed he is going ahead of you to Galilee; there you will see him.' This is my message for you." 8 So they left the tomb quickly with fear and great joy, and ran to tell his disciples. 9 Suddenly Jesus met them and said, "Greetings!" And they came to him, took hold of his feet, and worshiped him. 10 Then Jesus said to them, "Do not be afraid; go and tell my brothers to go to Galilee; there they will see me."

11 While they were going, some of the guard went into the city and told the chief priests everything that had happened. 12 After the priests had assembled with the elders, they devised a plan to give a large sum of money to the soldiers, 13 telling them, "You must say, 'His disciples came by night and stole him away while we were asleep.'" 14 If this comes to the governor's ears, we will satisfy him and keep you out of trouble." 15 So they took the money and did as they were directed. And this story is still told among the Jews to this day.

16 Now the eleven disciples went to Galilee, to the mountain to which Jesus had directed them. 17 When they saw him, they worshiped him; but some doubted. 18 And Jesus came and said to them, "All authority in heaven and on earth has been given to me. 19 Go therefore and make disciples of all nations, baptizing them in the name of the Father and of the Son and of the Holy Spirit, 20 and teaching them to obey everything that I have commanded you. And remember, I am with you always, to the end of the age."

c. A Day of Mutual Meeting

Along with Jews and Christians, Muslims believe in resurrection. Followers of Islam, called Muslims, embrace five fundamental pillars of faith: *shahadah* (Islamic creed), *salah* (daily prayer), *zakāt* (almsgiving), *sawm* (fasting during the month of Ramadam), and *hajj* (pilgrimage to Mecca at least once in a lifetime). They hold that Muhammad (or, as some spell, Mohammed or Mahomet) (610–32 CE) was the messenger of Allah or God who called people back to the pure faith delivered by

Abraham, Jesus, and the other prophets (Bilhartz 2006, 103). The sacred texts of Islam are the chapters in the Qur'an, which was revealed over a period of time to Muhammad in the seventh century CE. The Qur'an is divided into 114 chapters known as *surahs*. Surahs are arranged largely by length, with the longest appearing first and the shortest last (Bilhartz 2006, 113).

According to Muslim teaching, all humans will be resurrected on the "Day of Resurrection" (*Qiyamah*) or "Day of Mutual Meeting" and be called to account for their record on earth. After judgment comes the bliss of heaven or the punishment of hell (Bilhartz 2006, 146). As it is said, "Everyone shall taste death. And only on the Day of Resurrection shall you be paid your wages in full. And whoever is removed away from the Fire and admitted to Paradise, he indeed is successful. The life of this world is only the enjoyment of deception (a deceiving thing)" (*Noble Qur'an* 2011, Surah 3:185). The following passages from the Qur'an address the promise of resurrection embraced by Muslim believers.

Readings: (*Noble Qur'an* 2011, Surahs 3:185–94 and 40:15–20). (Note: *The Noble Qur'an* is considered a Sunni translation, which is widely disseminated and funded by the Saudi government. The Sunni tradition is one of the two major denominations in Islam. The Shia tradition is the second. The Sunnis are a majority in China, Africa, and South Asia. The Shias are a majority in Iran, Iraq, and Lebanon. Both Sunnis and Shias consider the Qur'an a sacred text. They differ on some matters concerning interpretation, succession of religious leadership, religious authority, customs, and holy days.)

[Surah 3:185–94]

185. Everyone shall taste death. And only on the Day of Resurrection shall you be paid your wages in full. And whoever is removed away from the Fire and admitted to Paradise, he indeed is successful. The life of this world is only the enjoyment of deception (a deceiving thing).

186. You shall certainly be tried and tested in your wealth and properties and in your personal selves, and you shall certainly hear much that will grieve you from those who received the Scripture before you (Jews and Christians) and from those who ascribe partners to Allah, but if you persevere patiently, and become *Al-Muttaqun* (the pious) then verily, that will be a determining factor in all affairs, and that is from the great matters, [which you must hold on with all your efforts].

187. (And remember) when Allah took a covenant from those who were given the Scripture (Jews and Christians) to make it (the news of the coming of

Prophet Muhammad and the religious knowledge) known and clear to mankind, and not to hide it, but they threw it away behind their backs, and purchased with it some miserable gain! And indeed worst is that which they bought.

188. Think not that those who rejoice in what they have done (or brought about), and love to be praised for what they have not done,—think not you that they are rescued from the torment, and for them is a painful torment.

189. And to Allah belongs the dominion of the heavens and the earth, and Allah has power over all things.

190. Verily! In the creation of the heavens and the earth, and in the alternation of night and day, there are indeed signs for men of understanding.

191. Those who remember Allah (always, and in prayers) standing, sitting, and lying down on their sides, and think deeply about the creation of the heavens and the earth, (saying): "Our Lord! You have not created (all) this without purpose, glory to You! (Exalted be You above all that they associate with You as partners). Give us salvation from the torment of the Fire.

192. "Our Lord! Verily, whom You admit to the Fire, indeed, You have disgraced him, and never will the *Zalimun* (polytheists and wrong-doers) find any helpers.

193. "Our Lord! Verily, we have heard the call of one (Muhammad) calling to Faith: "'Believe in your Lord,' and we have believed. Our Lord! Forgive us our sins and remit from us our evil deeds, and make us die in the state of righteousness along with *Al-Abrar* (those who are obedient to Allah and follow strictly His Orders).

194. "Our Lord! Grant us what You promised unto us through Your Messengers and disgrace us not on the Day of Resurrection, for You never break (Your) Promise."

[Surah 40:15–20]

15. (He is Allah) Owner of High Ranks and Degrees, the Owner of the Throne. He sends the Inspiration by His Command to any of His slaves He wills, that he (the person who receives inspiration) may warn (men) of the Day of Mutual Meeting (i.e. The Day of Resurrection).

16. The Day when they will (all) come out, nothing of them will be hidden from Allah. Whose is the kingdom this Day? (Allah Himself will reply to His Question): It is Allah's the One, the Irresistible!

17. This Day shall every person be recompensed for what he earned. No injustice (shall be done to anybody). Truly, Allah is Swift in reckoning.

18. And warn them (O Muhammad) of the Day that is drawing near (i.e. the Day of Resurrection), when the hearts will be choking the throats, and they can

neither return them (hearts) to their chests nor can they throw them out. There will be no friend, nor an intercessor for the *Zalimun* (polytheists and wrong-doers, etc.), who could be given heed to.

19. Allah knows the fraud of the eyes, and all that the breasts conceal.

20. And Allah judges with truth, while those to whom they invoke besides Him, cannot judge anything. Certainly, Allah! He is the All-Hearer, the All-Seer.

C. REFLECTIONS: WESTERN FUNERAL OR MEMORIAL SERVICE

1. Observe and analyze an Abrahamic funeral or memorial service: choose a particular tradition, go online, and view its funeral or memorial service. (You will be able to find one by googling, for instance, "Muslim funeral service video.") Explain what function the ritual serves in the tradition. Select at least *three* aspects of the service that relate to death and explain their meaning within the particular tradition. Be sure to give details in your analysis.

2.(*) Reflect further on how resurrection is understood in the readings and in any additional resurrectionist account you may wish to draw upon. In your own words, what is resurrection? Put another way, how would you explain resurrection to someone who is not familiar with the concept? Be sure to develop your thoughts on the matter and give detail.

3.(*) Do you believe that resurrection occurs? Why or why not? Be sure to develop your thoughts on the matter and give details.

D. FURTHER READINGS

Alexander, Eben. 2012. *Proof of Heaven: A Neurosurgeon's Journey into the Afterlife.* New York: Simon and Schuster.

Bilhartz, Terry D. 2006. *Sacred Words: A Source Book on the Great Religions of the World.* New York: McGraw Hill.

Chidester, David. 2002. *Patterns of Transcendence: Religion, Death, and Dying.* Belmont, CA: Wadsworth.

Madigen, Kevin J., and Jon D. Levenson. 2008. *Resurrection: The Power of God for Christians and Jews.* New Haven, CT: Yale University Press.

Rohde, Erwin. [1921] 1925. *Psyche: The Cult of Souls and Belief in Immortality among the Greeks*. New York: Harper and Row.

Smith, Huston. 1991. *The World's Religions*. New York: HarperSanFrancisco.

Tipler, Frank J. 1994. *The Physics of Immortality: Modern Cosmology, God and the Resurrection of the Dead*. New York: Doubleday.

— Chapter 6 —

Medical Immortality

A. CONTEXT

So far in our investigation of death, we have considered traditional ways of understanding death. The view that death is physical or psychological disintegration is a popular interpretation today because of the influence of science and medicine in how we understand our world. This influence leads us to hold that reality is composed of physical matter and that we come to know it through an empirical methodology. For instance, a brain is composed of cells, and we come to know it through physical examination and tests. In addition, death as reincarnation or resurrection is a long-standing interpretation of death supported by Asian and Western religious and spiritual views. As we have seen, accounts of reincarnation and resurrection rely on a view of reality that is not reducible to physical matter. They are based on the view that reality is (at least in part) transcendent and that we come to know it through a faith-based or nonempirical perspective.

In the twenty-first century, and with the rise of biomedical technology, additional ways of viewing death emerge. Today biomedical technology offers new methods to prolong life. Technology allows us to prolong life through lifesaving interventions, including the treatment of disease, thereby allowing patients to live longer than they otherwise would have lived (Benecke 2002). In prolonging life, medicine delays death, as evidenced in the change in the human life span in the last century. It is reported that the human life span increased from 49.2 years in 1900 to 76.9 years in 2000. By 2030, life expectancy is estimated to reach 82 or more years (Sprott 2008, 3). Perhaps someday the life span will reach hundreds, if not thousands, of years. With these increases in the human life span, a new view of immortality emerges, one made possible by biomedical inventions.

In addition to extending life, biomedical technologies prolong life, in a sense, when they develop cell lines that live on well past an individual's lived experience. Such is the story captured by Rebecca Skloot in *The Immortal Life of Henrietta Lacks* (2010). In this retelling of historical events, clinical researchers at Johns Hopkins removed cancer cells from Henrietta Lacks without her consent. Lacks's cells went on to make millions for researchers who developed "immortal" cell lines (known as the HeLa cells), which have been used to further our understanding of cancer cell growth. Twenty tons of cells and eleven thousand patents have been generated from such research, with little of the financial benefits going to the family. Recently, and as a result of responses to the story, the Lacks family has received financial compensation and recognition for their family member's significant contributions to our understanding of cancer.

For some, the Lacks's story illustrates that medical immortality through unregulated cell growth is not a pipe dream. Of course, while the Lacks's cell line may be used indefinitely, *immortality* may not be the best term to use for these cells because, in this case, we are not talking about the immortal life of a human being. Let's face it: there is much work to be done in developing the means just for an individual human to live for a very long time, through reengineering cellular division and the ability of cells to repair themselves, if that even becomes possible. Nevertheless, some claim that this kind of extended life span can, and for some *should*, be made available. This is the case because life is seen as a good, and more good is typically better than less good. This is also the case because some people would choose to live forever, for a multitude of reasons, such as meeting descendants and experiencing more of life. Yet opponents of efforts to extend the life span remind us that serious drawbacks to extending life need to be considered. For example, the extension of life does not always translate into quality of life, medicine can do harm as it attempts to extend life, and human rights may be jeopardized in an attempt to prolong life. The selections that follow are drawn from the works of British physician Kristie McNealy, British bioethicist John Harris, US bioethicist and attorney George Annas, and US cardiologist Anthony N. DeMaria. They provide a number of insights and concerns about attempts to seek medical immortality.

B. PERSPECTIVES

1. Immortality through Life Extension

Family Health Guide publishes daily news from well-regarded sources such as the *Journal of the American Medical Association*, the *British Medical Journal, Lancet,* the

British Medical Association, and leading universities, plus articles from its own editorial team. Physician Kristie McNealy serves on the editorial team and published a two-part article on the science behind life extension. In it, she reports, "Today, scientists are coming closer than ever to making real medical breakthroughs that will 'cure' aging and eventually bring an end to natural death. Pharmaceutical discoveries, and advances in the fields of nanotechnology, cloning, stem cell research and cryonics offer tantalizing glimpses at a future free from old age, and the ability to actually reverse the aging process itself—possibilities that life extension experts feel could become a reality by 2019" (2010, 1). In what follows, McNealy surveys some of the advantages and disadvantages of life extension made possible by biomedical technologies.

Reading: "In Pursuit of Immortality: The Science behind Life Extension" (McNealy 2010, 1–2)

Since the beginning of recorded history, humans have searched for the key to immortality and eternal youth. From the epic of Gilgamesh, recorded on clay tablets around 2000 B.C., to Ponce de Leon's famed search for the fountain of youth in the new world, the extension of life has been a recurring theme for humanity.

Today, scientists are coming closer than ever to making real medical breakthroughs that will "cure" aging and eventually bring an end to natural death. Pharmaceutical discoveries, and advances in the fields of nanotechnology, cloning, stem cell research and cryonics offer tantalizing glimpses at a future free from old age, and the ability to actually reverse the aging process itself—possibilities that life extension experts feel could become a reality by 2019. Of course, along with these discoveries come ethical questions about the meaning of life in the absence of death and the fate of religion, as well as concerns about overpopulation, boredom and why anyone would really want to live forever.

If the claims of life extension proponents sound far fetched, consider the fact that the average human lifespan has doubled since 1900 and continues to increase. Enormous medical advances occurred during the 20th century, resulting in the development of medications and technology that were once unthinkable. Less than 100 years ago, insulin was unknown and type 1 diabetes was a fatal and mysterious disease. Now, insulin is an inexpensive and easily obtainable drug that saves lives every day. Other medical devices that are common today, like internal pacemakers and contact lenses, were unthinkable just 100 years ago, and the rate of medical and scientific advances continues to increase.

In humans, like all mammals, aging begins almost as soon as we reach physical maturity. In fact, with an average lifespan of 70 years and maximum verified lifespan of 122 years, a typical person today spends the majority of their life aging. In biological terms, aging can be defined as an accumulation of damage to macromolecules like DNA, as well as cells, tissues and organs. The differences in life span between species are determined by genetics, differences in the ability to repair damaged DNA, as well as differences in antioxidant enzymes and free radical production. The science of life extension seeks to exploit these differences and alter the biological pathways involved in aging, with the ultimate goal of immortality. In this feature, we'll discuss some of the most promising approaches in life extension science.

Chemical Life Extension

For years it's been common knowledge that drinking red wine (in moderation) is good for your heart. What we didn't know was why. Then, in 1992, a chemical named resveratrol was suggested as the cause of the cardioprotective properties of wine. Unfortunately, you'd have to drink hundreds of bottles of wine each day to consume enough resveratrol to see a significant effect, so work on a pharmaceutical form of resveratrol was begun. Researchers have since shown that resveratrol extends the lifespan of yeast, but studies in other species have had conflicting results. A 2008 study on mice found that while resveratrol causes changes in gene expression that are similar to those caused by dietary restriction and mice had fewer signs of old age, unfortunately they didn't live any longer than typical mice. . . .

Therapeutic Cloning and Stem Cells

To some researchers, therapeutic cloning is the most practical approach to life extension because compared to other methods, current technology for cloning is much more advanced. We already know how to make stem cells, and are learning more constantly about how to direct them to form different types of cells. In order to clone the cells of a patient and create a new organ or tissue, the process starts by taking a donor egg, and replacing its nucleus with the nucleus from a patient's cell, creating an embryo. The new cell is then stimulated to divide, and after a few days, at the blastocyst stage of development, there are stem cells available to collect.

These brand new stem cells have the ability to turn into any type of cell in the human body, like muscle, blood cells or nerve cells. Scientists grow them in a culture dish, and are working to learn what chemicals or other instructions to add to tell them what type of organ or tissue to form. Ultimately, these newly formed, specialized cells will be injected into patients to heal damaged body parts, or used to grow whole new organs. One possible application of this process is to make bone marrow from stem cells and then inject them into the body. The bone marrow cells would find their way into the patient's bone marrow, and the new cells could rejuvenate many parts of the body. . . .

Nanotechnology

Some of the biggest advances in life extension may come from the microscopically small products of nanotechnology. Nanotechnology is the building of precise structures on an atomic or molecular scale. Ultimately, scientists hope to build molecular machines or robots that will revolutionize both medicine and manufacturing.

In the field of life extension, theoretical nanotech treatments include chromosome replacement therapy. In chromosome replacement therapy, a nanofactory would produce perfect new chromosomes based on the patient's own genome. Special nanorobots carrying these new chromosome sets would be injected into the patient. They would the travel to cells, extract the existing chromosomes and replace them with new copies, effectively eliminating age related DNA damage. . . .

Caloric Restriction

In order for humans alive today to see the full benefits of life extension science, we need to live long enough to see further advances in the field. Today, caloric restriction is the most well documented method for increasing lifespan, and it is the only dietary method that has been shown to improve the mean and maximum lifespan of a variety of species, from yeast to dogs. In fact, in rats, calorie restriction was shown to double life span 70 years ago.

Caloric restriction involves consuming a severely reduced calorie diet while still consuming adequate quantities of vitamins and minerals. Scientists are currently working to extend their knowledge about caloric restriction to primates. While we don't yet have proof that caloric restriction will increase the human lifespan, we do know that it lowers cholesterol, fasting glucose, and blood pressure. . . .

Cryonics/Cryogenics

While most people tend to avoid contemplating their own death, a few people are choosing to use cryonics or cryogenics to preserve their bodies after death. Although there are no guarantees that cryonics, the process of warming and then reviving and treating a patient whose body has been preserved at extremely cold temperatures, will ever become a reality, some feel that the cost of cryonics is a small price to pay for the chance to live again in the future.

The purpose behind cryonics is to slow metabolism and the process of decay using extremely low temperatures, in hopes that one day we will discover a way to treat the patient's cause of death. While people have been cryogenically frozen since the 1960s, no person or mammal has ever been revived from cryogenic temperatures. One of the many obstacles to successfully using cryogenics is learning how to repair the damage caused by ice crystals formed within cells. While improving freezing techniques to reduce cell damage is certainly a goal, researchers hope that in the future we will also have the ability to repair damage done by past and present freezing techniques.

Conclusion

Life extension proponents view aging as the ultimate disease and they aim to find a cure. They feel strongly that this cure will be discovered within our lifetime. However, life extension is not without controversy.

Many opponents of life extension approach the concept from a religious or philosophical perspective, worrying about the value of life without death, and how a lack of natural death would impact the true meaning of life. In many ways, this argument falls along religious lines, with people who believe they will be rewarded in the afterlife standing opposed to life extension. On the other hand, people who are less certain about what happens after death tend to see little harm in enjoying life for as long as possible, in case this life is all we get. Many on both sides question the religious implications of making humans immortal, and thus god-like.

More practical concerns about life extension include overpopulation. Life extension proponents point out that as life expectancy has increased, birth rate has naturally decreased. They also argue as population increases, humans become more efficient, allowing the world to sustain more inhabitants.

2. Immortality through Treating or Preventing Debilitating Illness

Sometimes good intentions carry unforeseen consequences. Such is a theme addressed by some in medicine about the use of medical interventions. Bioethicist John Harris holds the Sir David Alliance Chair of Bioethics at the University of Manchester in England. He is the author of numerous articles in bioethics. In "Intimations of Immortality," Harris addresses the implications of medical technologies and interventions for how they prolong human life and make possible forms of immortality. There are a number of challenges with such interventions, includ-ing access to such interventions and the harms that may result from them. What is noteworthy about Harris's contribution is his view that medicine's effort to treat disease may be seen to be efforts to prolong the life span. As he says: "We should remember that it [immortality] is connected with preventing or curing a whole range of serious diseases. It is one thing to ask whether we should increase people's life-spans, and to answer no; it is quite another to ask whether we should make people immune to heart disease, cancer, dementia, and to decide that we should not. It might thus be appropriate to think of 'immortality' as the, possibly unwanted, side effect of treating or preventing debilitating illness" (2000, 59). To critique medicine's efforts to prolong the life span is in some sense, then, to critique medicine's efforts in treating disease. But who among us would want medicine to stop treating disease? Who among us would call for an end to clinical interventions? In what follows, Harris entertains some of the challenges raised by medicine's efforts to prolong human life.

Reading: "Intimations of Immortality" (Harris 2000, 59). Reprinted with permis-sion from AAAS.

Is it simply a design fault that we age and die? If cells were not programmed to age; if the telomeres, which govern the number of times a cell divides, did not shorten with each division; if our bodies could repair damage due to disease and aging, we would live much longer and healthier lives. New research now allows a glimpse into a world in which aging—and even death—may no longer be in-evitable. Cloned human embryonic stem cells, appropriately reprogrammed, might be used for constant regeneration of organs and tissue. Injections of growth factors might put the body into a state of constant renewal. We may be able to switch off the genes in the early embryo that trigger aging, rendering it "immortal" (but not invulnerable). We do not know when, or even if, such tech-niques could be developed and made safe, but some scientists believe it is possible.

These scientific advances could lead to significantly extended life-spans, well beyond the maximum natural age of about 120 years. The development of these technologies may be far in the future, but the moral and social issues raised by them should be discussed now. Once a technology has been devel-oped, it may be difficult to stop or control. Equally, fears provoked by tech-nological developments may prove unfounded; acting precipitately on those fears may cut us off from real benefits. Scanning future horizons will enable us to choose and prepare for the futures that we want, or arm us against futures that, while undesired, we cannot prevent.

The technology required to enable extended life-spans is likely to be expensive. Increased life expectancy would therefore be confined, at least initially, to a small minority of the population even in technologically advanced countries. Globally, the divide between high-income and low-income countries would increase. Populations with increased life-spans would be unlike our aging populations. The new "immortals" would neither be old, nor frail, nor necessarily retired. We have, however, learned that ageism is a form of discrimination, and this may make it more difficult to resist the pressure for longevity.

We thus face the prospect of "mortals" and "immortals" existing alongside one another. Such parallel populations seem inherently undesirable, but it is not clear that we could, or should, do anything to prevent such a prospect for reasons of justice or morality. If increased life expectancy is a good, should we deny palpable goods to some people because we cannot provide them for everyone? We do not refuse kidney transplants to some patients because we cannot provide them for all, nor do we regard ourselves as wicked because we perform many such transplants, while low-income countries perform few or none at all.

Would substantially increased life expectancy be a benefit? Some people regard the prospect of "immortality" with distaste or even horror; others desire it above all else. Most people fear death, and the prospect of personal extended life-span is likely to be welcomed. But it is one thing to contemplate our own "immortality," quite another to contemplate a world in which increasing numbers of people live indefinitely, and in which future children have to compete with previous generations for jobs, space, and everything else.

Such a prospect may make "immortality" seem unattractive, but we should remember that it is connected with preventing or curing a whole range of serious diseases. It is one thing to ask whether we should increase people's life-spans, and to answer no; it is quite another to ask whether we should make people immune to heart disease, cancer, dementia, and to decide that we should not. It might thus be appropriate to think of "immortality" as the, possibly unwanted, side effect of treating or preventing debilitating illness.

3. Genetic Immortality

Today, genetic medicine makes possible clinical interventions previously unimagined. Think here of human cloning and germ line genetic engineering. Such clinical interventions raise numerous ethical issues. Attorney and bioethicist George J. Annas is the Edward R. Utley Professor and chair in the Department of Health Law, Bioethics, and Human Rights at the Boston University School of Public Health in Massachusetts. He has written on numerous topics in bioethics and the law. In what follows, Annas explores human cloning and inheritable genetic alterations from the perspective of the need to conserve the human species and the moral requirement to protect human rights. Annas proposes an international "Convention on the Preservation of the Human Species" that would outlaw all efforts to initiate a pregnancy using either intentionally modified genetic material or human replication cloning using somatic cell nuclear transfer. Here somatic cell nuclear transfer is a technique for creating a clone embryo with a donor somatic nucleus. As he concludes, "Whether in private laboratories and corporations or on the state, federal, or global level, the ethical and human rights challenge remains to protect the human rights of children by prohibiting their genetic manufacture while permitting legitimate research designed to make new medicine to proceed. We can, in short, pursue regenerative medicine without simultaneously pursuing replicative medicine, seeking immortality, or producing the posthuman" (2008, 34). In what follows, Annas explores the prospects and perils of immortality brought about by cloning and inheritable genetic alteration.

Reading: "Immortality through Cloning" (Annas 2008, 32–34). Reprinted with permission of Johns Hopkins University Press.

Biotechnology, especially human cloning and germline genetic engineering, have the potential to permit us to design our children and literally to change the characteristics of the human species. The movement toward a posthuman world can be characterized as a movement down a slippery slope to a neoeugenics that will result in the creation of one or more subspecies or superspecies of humans. The first vision sees science as our guide and ultimate goal. The second is more based on our human history, which has consistently emphasized differences and has used those differences to justify genocidal actions. It is the propect of "genetic genocide" that calls for treating cloning and genetic engineering as potential weapons of mass destruction and the unaccountable genetic researcher as a potential bioterrorist.

   The greatest accomplishment of humans has not been our science but our development of human rights and democracy. Science deals with facts, not values. It cannot, for example, tell us whether immortality, even limited to genetic immortality, is good for either individuals or the human species. Because science cannot tell us what we should do, or even what our goals are, humans must give direction to science. In the area of genetics, this calls for international action to control the techniques that could lead us to commit species suicide. We humans recognized that risk clearly in splitting the atom and developing nuclear weapons, and most humans recognize the risk in using human genes to modify ourselves. Because the risk is to the entire species, it requires a species response. Many countries have already enacted bans, moratoria, and strict regulations on various species-altering technologies. The challenge, however, remains global, and action on the international level is required to be effective.

One action called for today is the ratification of an International Convention for the Preservation of the Human Species that outlaws human cloning and inheritable genetic alterations, or at least the latter. This ban would not only be important for itself but would also mark the first time the world worked together to control a biotechnology. Cloning and inheritable genetic modifications are not bioweapons per se, but they could prove just as destructive to the human species if left to the market and to individual wants and desires. An international consensus to ban these technologies already exists, and countries, NGOs, and individual citizens should actively support a renewed treaty process, as they did with the recent Convention on the Prohibition of the Use, Stockpiling, Production and Transfer of Anti-personnel Mines and their Destruction (Mine Ban Treaty).

Inheritable genetic alteration of children may not seem as important as land mines because no child with such attempted alterations has yet been born and thus no child has yet been harmed by this technique. Nonetheless, the use of inheritable genetic alteration (and human replicative cloning) has the potential to harm all children, both directly by limiting the genetically modified children's freedom and harming them physically and mentally, and indirectly by devaluing all children by treating them as products of their parents' genetic specifications. Of more concern, inheritable genetic alteration carries the prospect of developing new species or subspecies of humans, who could turn into either destroyers or victims of the human species. Opposition to cloning and inheritable genetic alteration is conservative in the strict sense of the word: it seeks to conserve the human species. But it is also liberal in the strict sense of the word: it seeks to preserve democracy, freedom, and universal human rights for all members of the human species.

Proponents of going full speed ahead with inheritable genetic alterations are fond of quoting Thucydides in his *History of the Peloponnesian War* to the effect that "the bravest are surely those with the clearest vision of the future, disaster and benefit alike, and that not withstanding these possibilities they move ahead." Sounds great, but when placed in historical context, the statement actually supports application of the precautionary principle to inheritable genetic alterations. The statement is made by the great Athenian general, Pericles, in his famous funeral oration praising the nature of Athenian citizenship. His real point is not that Athenians are impulsive and brave as a character trait, but rather that as a democracy they think before they act and weigh the possible consequences before voting on how to proceed. As Pericles puts it concisely, "The worst thing is to rush into action before the consequences have been properly debated." And later, when Athens is in the midst of a disastrous war, Thucydides, a former general himself, describes how war changes what we think of as virtue: "What used to be described as a thoughtless act of aggression was now regarded as courage. . . . To think of the future and wait was merely another way of saying one was a coward. . . . Ability to understand a question from all sides meant that one was totally unfitted for action. Fanatical enthusiasm was the mark of a real man."

Whether in private laboratories and corporations or on the state, federal, or global level, the ethical and human rights challenge remains to protect the human rights of children by prohibiting their genetic manufacture while permitting legitimate research designed to make new medicine to proceed. We can, in short, pursue regenerative medicine without simultaneously pursuing replicative medicine, seeking immortality, or producing the posthuman.

4. Immortality through a Cell Line

Cardiologist Anthony N. DeMaria shares with Annas concerns about attempts to achieve medical immortality through biomedical technologies. DeMaria is the editor-in-chief of the *Journal of the American College of Cardiology*, located in San Diego, California. In "Problems with Immortality," DeMaria addresses the advantages and disadvantages of achieving immortality through one's cell line. He references Henrietta Lacks, whose cells were taken by her doctor George Gey at Johns Hopkins Hospital in 1951 without her consent. The cells, which have come to be known as the HeLa cells, lived to develop into a cell line that has helped millions of patients. One of the results of the Lacks case is a discussion of the extent to which the cells in our bodies are the property of the living. Although the US Supreme Court has ruled that discarded cells in the clinical setting are not the property of the patient or research subject (see the 1990 case *Moore v. Regents of the State of California*), the matter in general is anything but settled, especially because financial interests are at stake. And the matter of medical immortality is anything but simple. As DeMaria says, "Although immortality remains an elusive and presumably extremely desirable goal, it is clearly not without its problems" (2010, ¶9). What follows is a look at a challenging topic made possible by new biomedical technologies.

Reading: "Problems with Immortality" (DeMaria 2010, 2140–42). Copyright Elsevier 2010. Reprinted with the permission of Elsevier. Address correspondence to: Dr. Anthony N. DeMaria, Editor-in-Chief, *Journal of the American College of Cardiology*, 3655 Nobel Drive, Suite 630, San Diego, California 92112, American College of Cardiology Foundation.

Although immortality is, I guess, one of the ultimate goals of medicine, no human has thus far achieved it, nor is anyone likely to do so for the foreseeable future. However, the same is not true for human tissue. I recently read a book about Henrietta Lacks [Skloot 2010], the woman whose cancer provided the HeLa cell line that has thus far proved immortal and has been responsible for many fundamental scientific breakthroughs. Although these cells and the discoveries they have enabled are a cause for celebration, they have also been representative of a number of problematic issues in medicine, many of which remain unresolved. In fact, some of the issues may be even more prominent today as we strive to develop stem cells and achieve tissue regeneration.

    Henrietta Lacks was a poor African-American woman living in Baltimore who developed a very malignant carcinoma of the cervix at a young age. The cells from this tumor were taken, for all intents and purposes without her knowledge or consent, and sent for possible tissue culture at Johns Hopkins University. They proved to be the first "immortal" human cells capable of forever continuously reproducing when cultured outside of the body, enabled many medical advances, and also became a profitable commercial product. Today, HeLa cells can be found in laboratories throughout the world, and they continue to be the cornerstone of a large body of medical research.

    The first aspect of the HeLa story that struck me, as it might most clinical investigators, was the serendipity of discovery. George Gey, the Hopkins researcher who grew the cells, had been unsuccessfully trying to culture human cells for many years. He had experimented with a myriad of culture media recipes without success. In fact, there was nothing unique or special about the media in which he placed Henrietta Lacks' cells in comparison to that used for many other cells that did not reproduce. The "discovery" was the result of the good fortune of obtaining cells that were almost indestructible. It is amazing to think of how many critical medical discoveries have depended heavily upon chance. However, it is also true that luck leads to important discoveries when it encounters the prepared mind. Henrietta's cells were absolutely unique, but they would have gone undetected and unrecognized had not George Gey been looking and been irrepressible in his search.

A perhaps weightier issue regarding the HeLa story relates to the socioeconomic status of the donor. Henrietta Lacks was a poor woman with little formal education who came under the clinical care of a research medical institution that provided safety net treatment for the uninsured. Although less prevalent today, this arrangement was fairly typical of the times (1950s to 1960s). So, the lower socioeconomic classes typically contributed disproportionately to the pool of research patients. Patients were sometimes uninformed about investigation, much less asked for their consent. We have long since rejected any such notions, and many of the courageous and altruistic patients who participate at uncertain risk in clinical trials now come from middle and upper socioeconomic classes. Nevertheless, this was probably not a proud moment in medical history, and today there still remains a bit of a divide in early stage research participation between patients and physicians in private hospitals and those in academic medical centers.

Neither Henrietta nor her family knew that her cells had been taken for research and were financially valuable until some time after her death. The issue of informed consent has now been well addressed in the United States. In fact, some would say it has perhaps gone too far. Approval is now often required for activities that could not constitute either risk or financial benefit to the patient. Nevertheless, it is probably better to do too much than too little. It is startling, however, to see the change in attitudes in the last several decades, and to realize that only a short time ago research was performed without the knowledge or understanding of the subjects.

Perhaps the most problematic, and still largely unresolved, issue raised by the HeLa story concerns the ownership of tissue once it is removed from the body. Selling Henrietta's cells ultimately provided a profitable business opportunity although Dr. Gey and Hopkins gave the cells away for free. However, neither Henrietta nor any of her relatives have ever received a penny of income. Among the questions raised is whether an individual has to be informed that tissue removed from his or her body may be used for research and has to consent to this use; the answer is apparently no. I must admit, I have had blood drawn for tests many times, but I have never wondered where the specimen has gone, and certainly never considered that it might be used for research. But, it appears that is entirely possible; we legally lose all ownership of our body parts whenever they are removed from the whole. The ancillary issue is who should benefit if the tissue leads to a profitable commercial business. At the moment, the answer seems to be that the patient has neither a right nor recourse to any financial gain from the tissue that he or she provided for the research.

The above issues of informed consent and financial gain are controversial and are the subjects of considerable ongoing debate. As a patient, I instinctively feel that, if it is my body, I should own the part, should certainly be informed if and how it will be used, and should share in any financial gain. This certainly seems to be true for all of our other worldly possessions. Although it is logical that we have a strong responsibility to do what is best for society with our tissue, this does not apply to other possessions such as money or property. Those opposed to this concept (not surprisingly, primarily the scientific and medical industry community) contend that the general rules governing the conduct of clinical research already provide adequate protection for the rights of patients regarding their tissues. Importantly, they argue that our moral obligation to benefit society outweighs personal considerations, and that good health differs from other possessions in this regard. They point out the logistical difficulties in trying to give patients control of their tissues. The actual investigation the specimen is used for may occur many years after and for a problem unforeseen at the time of donation. Finally, those opposed to patient control worry that the potential of a financial benefit may lead to unrealistic demands and negotiations and, eventually, to the loss of important opportunities for medical advances. Of course, this applies equally well to the intellectual property rights that the scientific community values and utilizes.

While the issues relating to the control of removed biological material have always existed, they may be even more pressing in the future. Cardiac regeneration utilizing stem cells is one of the major thrusts of contemporary research. We have participated in several studies in which the evidence of immune protection of stem cells has been exploited to utilize donor cells for therapy. Prior to reading Skloot's book, I would not have thought to question the consent of the donors or their participation in any profits. In addition, the revolution in genetics/genomics has raised questions regarding the ownership of genes and genetic material. It seems clear that these questions will receive more rather than less attention as time goes by.

Rebecca Skloot has written a magnificent book, and one which raises important questions for us in medicine to address. We are walked through the serendipitous discovery of "immortal" cells at Hopkins, and we wonder how many other similar tissues were discarded due to the lack of interest of those in charge. As the saying goes, those who work the hardest often have the most luck. We are once again reminded of the crucial importance of our patients to our research; they are the real heroes. As for the ownership of biological material removed from the body, I am ambivalent. It seems absolutely reasonable that patients should have a say in whether and how their tissues might be used, and share in any profits that might result. While it might be an unearned gift to have tissue of great value, the same could be said for great intelligence, or a great voice or athletic ability, and so on. On the other hand, if a tissue had the potential to provide a cure for cancer, I cannot see how the rights of the patient could outweigh the potential benefit to society. We have given government the right of eminent domain to acquire real estate, so surely we could do the same for biological material. As in all things in medicine, we as physicians should take the lead in resolving these dilemmas. Although immortality remains an elusive and presumably extremely desirable goal, it is clearly not without its problems.

C. REFLECTIONS: EXTENDING THE LIFE SPAN AND THE COST OF DYING

1.(*) Pursue a point raised by Kristie McNealy: Do you think a lack of natural death in human existence would affect the true meaning of life? Put another way, if we did not die, would this change the meaning of life? Why or why not? Be sure to develop your thoughts on the matter.

2.(*) If you were offered medical immortality, would you choose it? Why or why not? Develop your reasoning in light of the discussions and with detail.

3. This is a continuation of your advance planning in preparation for death and concerns the cost of dying. An article in the *Wall Street Journal* (Rockoff 2014) provides the following statistics about end-of-life health care costs in the United States:

- $50 Billion: Annual spending, about 10 percent of Medicare funds, on the last month of a patient's life.
- 65 percent: The share of health care spending that goes toward the sickest 10 percent of patients, an average of $157,510 annually per patient.
- 65 percent: The share of poor-prognosis cancer patients who are hospitalized during the last month of life.
- 25 percent: The share of poor-prognosis cancer patients who use a hospital intensive care unit during the last month of life.
- 30 percent: Percentage of poor-prognosis cancer patients who die in the hospital.

- 54 percent: The share of cancer patients who use hospice during the last month of life.
- $5,000 to $7,000: The annual savings per patient when palliative care is provided alongside usual care.

The point is this: it's expensive to die. Your task is to come up with a written estimate of the cost of your care at the end of life. To do this, decide how you will die. Will you die young with an acute illness or sudden accident, or will you die old and with chronic care? (Reflect on what is likely given your lifestyle, your medical history, and your family history.) Price out how much it will be for you to die in this situation. Some things to price out, depending on your situation, include:

a. hospital care
b. hospice care
c. long-term care in a nursing home
d. private care in your home
e. ambulance service
f. medications
g. other costs (renovations to your home or car, cleaning service, food preparation, special aids, wheelchair, walker, cane, oxygen, shower chair)

In your answer, list your expenses, with their cost, and add up the costs for a total. (If you have insurance, still price out how much it will cost to die.) What do your expenses tell you about what is important to you when you die?

D. FURTHER READINGS

Barry, Vincent. 2007. *Philosophical Thinking about Death and Dying*. Belmont, CA: Thomson Wadsworth.

Benecke, Mark. 2002. *The Dream of Eternal Life: Biomedicine, Aging, and Immortality*. New York: Columbia University Press.

Bhattacharya, Pranab Kumar. 2013. "Is There Science behind the Near-Death Experience: Does Human Consciousness Survive after Death?" *Annals of Tropical Medicine and Public Health* 6 (2): 151–65.

Bova, Ben. 1998. *Immortality: How Science Is Extending Your Life Span—and Changing the World*. New York: Avon Books.

Cave, Stephen. 2012. *Immortality: The Quest to Live Forever and How It Drives Civilization*. New York: Crown.

Devine, Claire. 2010. "Tissue Rights and Ownership: Is a Cell Clone a Research Tool or a Person?" *Columbia Science and Technology Law Review* 29 (March 9). www.stlr.org/2010/03/tissue-rights-and-ownership-is-a-cell-line-a-research-tool-or-a-person/.

Harris, John. 2007. *Enhancing Evolution: The Ethical Case for Making People Better*. Princeton, NJ: Princeton University Press.

Moore v. Regents of the University of California. 1990. 51 Cal. 3d 120; 271 Cal. Rptr. 146; 793 P.2d 479.

Parry, Bronwyn. 2004. "Technologies of Immortality: Brain on Ice." *Studies in History and Philosophy of Biological and Biomedical Sciences* 35 (2): 391–94. http://dx.doi.org/10.1016/j.bbr.2011.03.031.

Rockloff, Jonathan D. 2014. "Palliative Care Gains Favor as It Lowers Costs." *Wall Street Journal*, February 23. www.wsj.com/articles/SB10001424052702303942404579363050214972722.

Skloot, Rebecca. 2010. *The Immortal Life of Henrietta Lacks*. New York: Crown.

Sprott, Richard L. 2008. "Reality Check: What Is Genetic Research on Aging Likely to Produce, and What Are the Ethical and Clinical Implications of Those Advances?" In *Aging, Biotechnology, and the Future*, edited by Catherine Y. Read, Robert C. Green, and Michael A. Smyer, 3–9. Baltimore: Johns Hopkins University Press.

Chapter 7

Digital Immortality

A. CONTEXT

With the rise of digital technology and specifically thanatechnology (digital technology that can be used to deal with death, dying, grief, loss, and illness) in the late
twentieth century, new possibilities such as "digital immortality" arise. *Digital immortality* refers to life after death through digital technology and has various meanings. Computer scientists Gordon Bell and Jim Gray understand digital immortality
in terms of a continuum "from enduring fame at one end to endless experience and
learning at the other, stopping just short of endless life" (2001, 29). Others understand digital immortality to include "endless life" in cyberspace. Putting the idea
provocatively, as philosopher Richard Jones does, digital immortality involves getting one's "mind out of one's brain" (2016, 183). Getting one's mind out of one's
brain is currently made possible through social networking sites, such as Facebook,
Twitter, and the like, and the ability to post "who I am," which endures well after
one's death. Attention has recently turned to how social networking sites can become a form of memorial after one's death and how emergent technologies can define new forms of immortality and afterlife (see the website Death and Digital
Legacy, at deathanddigitallegacy.com). As *Time* reporter Gaelle Faure comments, "As
people spend more time at keyboards, there's less being stored away in dusty attics
for family and friends to hang on to. . . . The pieces of our lives that we put online
can feel as eternal as the Internet itself" (2009, 1).

What follows is a selection of readings that raise various aspects of a new way to
understand death: death as digital immortality. Some support digital immortality,

and others are gravely concerned about any attempt to seek immortality in the digital world. Selections are drawn from the writings of journalist Chris Faraone, philosopher Richard Jones, philosopher Patrick Stokes, and grief counselors Carla Sofka and her colleagues.

B. PERSPECTIVES

1. Digital Death

The topic of digital death emerges in widely available publications. Journalist Chris Faraone has written for a number of publications, including the *Boston Phoenix*, *Boston Herald*, *Fast Company*, *Spin*, the *Source*, the *Columbia Journalism Review*, and *digboston*. While at the *Boston Phoenix*, he wrote a version of this piece on digital death, which was picked up by the Colorado Springs newspaper, the *Independent*, revised, and reprinted. In the article, Faraone addresses the need to think further about digital death and where data goes after one's death ("Chris Faraone" 2017). The boom in the digital death industry draws our attention to the challenges that we may encounter if we do not think about what will happen to our data in cyberspace. As Faraone says:

> Even mega networks have policies that experts say are inadequate for the age we live in; Facebook, which loses three users every minute to the Farmville in the sky, has yet to implement a setting that allows users to decide their page's fate. According to the Digital Beyond, the go-to forum for e-death news, Twitter makes the bereavement process easy—all they require is a link to a public obit or news article and they'll promptly delete an account; they'll even provide mourners with a backup.
>
> On the other hand, Yahoo!, which owns Flickr and other popular services, buries "no right to survivorship" and "non-transferability" clauses in their sign-up-box legalese. Apple deals with the deceased on a case-by-case basis, but don't expect to inherit your uncle's iTunes library. Until a court decides differently, that's copyright infringement. (2011, 17)

Although some of these policies have changed since Faraone penned this, the moral of the story is that the digital age presents some new challenges in planning for one's death. One has an opportunity, then, to think about what will happen to one's digital information. If one does not plan for one's death in this digital day-and-age, one may find immortal existence in cyberspace, whether or not one chooses it. As Faraone says, "*My status may be 'I'm Dead,' but I'll be on Facebook forever*" (2011, 17). The essay that follows draws our attention to the timely topic of "Where does one's data go after one's death?"

Reading: "Digital Death: Where Does Your Data Go When You Reach the End of the Road?" (Faraone 2011)

I don't plan to live past 50. If I keep this pedal pinned to the floor, even that might be pushing it. Death is something that I ponder daily, usually between my morning blunt and Burger King breakfast run.

Yet for all my morbid musings, I've never thought much about my digital legacy—the only significant asset that I have. In addition to all the articles I've written that exist online, I've got more photos, profiles and social-networking accounts than the average Web junkie, and a whole lot of enemies who would flame my wall in the event of my demise. I don't care about my body; like comedian David Cross, I'm donating my dead meat to necrophiliacs (if possible; there's no check box for that). It's my virtual soul that I wish to preserve.

What would happen if I logged off for good, and took my passwords to the grave? Will spambots devour my blogs even as maggots chomp my corpse?

"You're delusional if you think everything you put on the Web is going to be there forever," said Adele McAlear, who founded the site Death and Digital Legacy, at a recent lecture I attended at SXSW. Her statement shook me. A top social-media consultant and seasoned lecturer on these matters, McLear says the U.S. Supreme Court will inevitably have to determine how companies handle data belonging to dead users.

In the absence of meaningful state or federal regulation regarding post-mortem rights in cyberspace, a so-called digital death industry is booming. Sites like DataInherit, a "Swiss bank for data," have tens of thousands of users in 100-plus countries. MemoryOf, which allows survivors to build tribute pages, is approaching the 100,000-memorial mark, while the comparable 1000Memories recently became the first company of its kind to attract a seven-figure capital injection.

There's been some notable coverage of digital death; business publications are especially enthralled by the potential of this relatively uninhabited marketplace. But I chose a more practical approach to probe the phenomenon, and decided to write out my own Web will.

Counting the Cloud

To that end, of the dozen or so services I could have used for digital asset man-agement, I picked the Madison, Wis.–based Entrustet, which has emerged as an industry leader. The three-year-old company's co-founder, Jesse Davis, is a young healthy dude who I think will be around for a while, and who says his com-pany has taken serious safety precautions, going "above and beyond" standard security measures to deeply encrypt information.

The first thing Davis advises me to do is "cloud count," or take an inventory of every site and service I belong to. Aside from the basics—Twitter, YouTube, Gmail, Tumblr, Facebook, and an interminable MySpace—there are several other accounts that I want closed, or at least maintained, after I pass. There's the eBay profile that I use to sell old comics for beer money, and the Adult Friend Finder account from my truly degenerate days. (In some sort of sick metaphor, hookup-site memberships are a major pain to get rid of.) I also have a few Word-Press blogs, SpringPad for my field notes, and online Bank of America access.

After I die, my relatives can contact these companies directly, and follow procedures to get into my accounts. But the process can be difficult. Even mega networks have policies that experts say are inadequate for the age we live in; Facebook, which loses three users every minute to the Farmville in the sky, has yet to implement a setting that allows users to decide their page's fate. Accord-ing to the Digital Beyond, the go-to forum for e-death news, Twitter makes the bereavement process easy—all they require is a link to a public obit or news ar-ticle and they'll promptly delete an account; they'll even provide mourners with a backup.

On the other hand, Yahoo!, which owns Flickr and other popular services, buries "no right to survivorship" and "non-transferability" clauses in their sign-up-box legalese. Apple deals with the deceased on a case-by-case basis, but don't expect to inherit your uncle's iTunes library. Until a court decides differently, that's copyright infringement.

Executor Privilege

To make my eventual dirt nap easy for everyone around me, I had to choose a digital executor to entrust with all of my passwords and uploaded files. I could have gone with mom or dad, though I don't think either would agree to run my Tumblr if I croak. None of my close guy friends are responsible enough to tackle such a major task, and as for my girlfriend—I considered it, but there's no safe-guard to protect me in the case that I'm executed by my executor.

I finally got my editor to agree to Entrustet's e-mail asking him to serve as my keeper. It was time to make decisions regarding my digital properties. For everything from Gmail to Facebook, Entrustet prompts me to enter my user name and password, and to choose an appropriate action to carry out in case I drop. The basic service is free, and allows me to transfer accounts to my "heir," who can then follow my instructions to curate or delete my profiles. For $30 a year I can upgrade to premium, which permits Entrustet to "incinerate" accounts I want terminated, wiping all traces of their existence from the Internets.

As a bonus, Entrustet allows me to upload documents for my executor's eyes only. Since most of my work is already online, I have just a few files for Carly to manage: some chapters of my books-in-progress on hip-hop and the Red Sox; a short story I've been trying to sell to McSweeney's for a decade; an autobiographic screenplay that I'd like hand-delivered to Darren Aronofsky. I'm also including a short list of songs to play at my wake: "Dead Letter" by Royal Flush, Biggie's "You're Nobody 'Til Somebody Kills You," and "Dead Body Disposal" by Necro.

Finally, I turned to another service, the relatively simple If I Die (ifidie.org), to assist with delivering goodbye salutations to my friends and family. With a few keystrokes, I sign up to store farewell notes, which will be sent to designated e-mail addresses if administrators determine, through various checkpoints, that I have kicked the bucket.

At first I planned to agonize over virtual daps to everybody from my mom to my high-school English teacher. But that exercise triggered an awful panic attack—sweats and shakes—so instead I went with something more generic. You might say it's the last auto-response message that I'll ever write. It's something like:

Subject: "Out of office until . . ."

If you're reading this e-mail, it means I either died, or found out where Tupac was hiding. Please don't cry for me, though—I had a kickass life, from meeting Daniel Ellsberg and every member of the Wu-Tang Clan, to toasting the 100th anniversary of legendary Boston bar J.J. Foley's with Jerry Foley himself.

I went to more than a few ballgames and concerts, and had friends who loved me back. Somehow I got a master's degree—almost two of them—and one time I even changed a car tire. Christopher Hitchens once stole a joke from me without knowing that I stole it from an episode of Dr. Katz.

I'm sad to say that I don't have any children or sperm that's banked somewhere. However, if you look around my futon, you'll probably be able to scrape up a little something. And if you have luck engineering my offspring, please tell the kid that he or she can read about me on the Internet, where my legacy will live on. My status may be "I'm Dead," but I'll be on Facebook forever.

2. Technological Immortality

Philosophers have also expressed interest in how immortality can be achieved through digital technology. Philosopher Richard A. Jones has been writing on this topic for some time now. Jones has taught philosophy at Howard University in Washington, D.C.; has been co-coordinator of the Radical Philosophy Association; and has published articles on topics concerning black studies, Wittgenstein and race, and postmodern racial dialectics. In a number of his writings, Jones explores how immortality can be achieved through digital technology. He shows that, with the aid of computer science, human beings will be able to achieve immortality, but that immortality will not be the theistic promise of the resurrection of the body and the soul. Rather, it will involve transferring the mind from one body to another. As Jones says, "The popular imagination is awash in cinematographic imagery of transpersonal identities, survival of the soul, and life after death. Films like *Being John Malkovich, Vanilla Sky, 2001: A Space Odyssey, Brainstorm,* and *The Seven Grams,* to name but a few, lend credence to the ideal that minds can be transferred from one body to another. These films whet the appetite for immortality, but the techniques for 'capturing' the mind/soul on magnetic substrates comes from computer science" (2016, 183). In what follows, Jones asks us to contemplate whether we could—or should—support transferring a mind from one body to another for purposes of achieving immortality.

Reading: "The Technology of Immortality, the Soul, and Human Identity," in *Postmodern Racial Dialectics: Philosophy beyond the Pale* (Jones 2016, 183–85). Reprinted with the permission of Rowan and Littlefield.

The popular imagination is awash in cinematographic imagery of transpersonal identities, survival of the soul, and life after death. Films like *Being John Malkovich, Vanilla Sky, 2001: A Space Odyssey, Brainstorm,* and *The Seven Grams,* to name but a few, lend credence to the ideal that minds can be transferred from one body to another. These films whet the appetite for immortality, but the techniques for "capturing" the mind/soul on magnetic substrates comes from computer science.

The Technology of Immortality

In 1988 Hans Moravec, then at the MIT Media Labs, wrote a seminal book entitled *Mind Children: The Future of Robot and Human Intelligence*. Moravec argues that physical death is the one problem that has caused more human misery and worry than any other. He further suggests that not only is solving this
problem within the technological capabilities of humankind, but that this is also
its ultimate teleological purpose. As a first stage, Moravec sees no difficulty with
the "transmigration" of human consciousness to more durable robots. The movies *Bladerunner*, *I Robot*, and *AI* immediately come to mind. He writes, "So
what about replacing everything, that is, transplanting the human brain into a
specially designed robot body? Unfortunately, while this solution might overcome most of our physical limitations, it would leave untouched our biggest
handicap, the limited and fixed intelligence of the human brain. This transplant
scenario gets our brain out of our body. Is there a way to get our mind out of our
brain?"

Moravec argues that the threshold density for computer storage techniques
is just a few generations from having the capacity to record the contents of
consciousness. This "snapshot" of consciousness could then be downloaded
onto computer software, where it could be stored (and/or augmented). Once a
year, like renewing a driver's license, one would store a copy of consciousness. Of course, critical consciousnesses (e.g., creative geniuses and political leaders) would have their consciousnesses scanned daily; this back-up procedure would protect against lost cogno-bytes. Real-time scanning of human
consciousness for storage and retrieval would then only be a few innovations
removed. Dave Egger's recent best-selling novel *The Circle* provides an easily
imaginable scenario where instantaneous data-capture of people's lives runs
amok. Storage for the billions of neuronal energy patterns would require ho-
lographic, laser-encoded, crystal-optic, molecular quantum-level storage techniques (HLCOMQST). But the quantum addressing, storage and retrieval techniques are currently within the imagination's production, if not technological
production. Moravec writes:

> Though you have not lost consciousness, or even your train of thought, your
> mind has been removed from the brain and transferred to a machine. In a
> final, disorienting step the surgeon lifts out his hand. Your suddenly aban-
> doned body goes into spasms and dies. For a moment you experience only
> quiet and dark. Then, once again, you can open your eyes. Your perspective
> has shifted. The computer simulation has been disconnected from the cable

leading to the surgeon's hand and reconnected to a shiny new body of the style, color, and material of your choice. Your metamorphosis is complete.

Moravec believes that your children's children will be perhaps the last generation to face the prospects of physical death. Thus freed from fear, post-humans will be enabled to complete sentiences' work in the universe, which is to "resurrect" all the sentient beings that have ever existed. Even without "physical" brains, the mathematical patterns of their contents, encoded in computer memory, will remain and interact with "mind's children." . . .

It is not only the transference of human memory and personality that Moravec envisions. Cosmologists John Barrow and Frank Tipler also agree. They write [in their book *The Anthropic Cosmological Principle*], that a "*Final Anthropic Principle*" (FAP): *Intelligent information-processing must come into existence in the Universe, and, once it comes into existence, it will never die out.*" Barrow and Tipler's Anthropic Principle requires artificial life as part of an evolutionary teleological end for the universe to completely cognize itself. Thus Moravec's technological storage and retrieval of human memory and the computer software necessary to simulate an ongoing existence is more than the "jacking-in" of human consciousness into *Neuromancer*'s "Cyberspace." In fact, in my estimation like the Final Anthropic Principle, Moravec's position is highly teleological: the purpose of intelligence in the universe is to transform the physical universe itself into an intelligent universe.

Computer technology is a hardware means to encode the "wet-ware" of human brainstate consciousness into a software, which in turn can be transformed from a continuous analog process into a digital format. The Rapid Fourier-Transforms necessary to reduce the complex variations of human neural electromagnetic wave output activity onto a DVD-like substrate and store it in computer memory is on the technological horizon. The recording of the soul therefore need not be beyond logical possibility. In the concluding chapter of *The Anthropic Cosmological Principle*, "The Future of the Universe," Barrow and Tipler write:

> An intelligent being—or more generously, any living creature—is fundamentally a type of computer, and is thus subject to the limitations imposed on computers by the laws of physics. However, the really important part of a computer is not the particular hardware, but the program; we may even say that a human being *is* a program designed to run on a particular hardware called a human body, coding its data in a very special type of data storage devices called DNA molecules and nerve cells. The essence of a human

being is not the body but the program which controls the body; we might even identify the program which controls the body with the religious notion of a *soul*, for both are defined to be non-material entities which are the essence of a human personality. In fact, defining the soul to be a type of program has much in common with Aristotle and Aquinas' definition of the soul as "the form of activity of the body."

In light of Moravec's contention that technological immortality is only a few generations of innovation in computer storage and retrieval away, the question remains: even if we could, why would we want to do this? Would it be ethical, and would it make us less human?

3. Facebook Immortality

Patrick Stokes is another philosopher who has interest in questions concerning digital immortality. Stokes is a lecturer in philosophy at Deakin University in Melbourne, Australia. His scholarship focuses on personal identity, narrative selfhood, and the metaphysics of death ("Patrick Stokes" 2017). In what follows, Stokes addresses whether the dead can live on in the digital technology of Facebook. His answer is "yes" and "no." As he says,

> Persons can, in some dimensions of their identity, survive their deaths in ways that selves cannot. Selves are rooted in first-personal, present-tense experience in ways that are quite different to persons, and this allows persons to persist in circumstances where the associated self has dropped out of existence. So while from my perspective the people I've lost to death survive in some tragically reduced and ontologically ambiguous form, from their own perspective they do not survive their death at all—because their death is precisely the loss of the self that constitutes their own perspective. One can only survive one's death second- or third-personally, so from a first-personal perspective, no one else is any better off than me. (2012, 377)

In other words, while one may live on in the memory of another, one cannot live on in one's own memory after death. Thus Facebook provides an immortal existence for oneself through others but not through oneself. This is the case because, once dead, one cannot experience oneself online. A consequence is that there is no sense of personal or "first-person" immortality that can be achieved through the digital technology of Facebook; there is only "third-person" immortality.

Reading: "Ghosts in the Machine: Do the Dead Live On in Facebook?" (Stokes 2012, 363–64, 377–78). Reprinted with the permission of Springer.

Every community has to find ways to deal with the deaths of its members, and since their advent, online communities have been no exception. From suicides streamed via webcam (Thompson 2008), to bloggers and social network users faking their deaths in apparent cases of "Munchausen by Internet" (Feldman 2000; Swains 2007, 2009), from websites that send out pre-prepared emails in the event of a user's death to ensure their passwords, financial details and darkest secrets do not die with them, to fake celebrity death rumours that sweep across the internet like wildfire, online life and offline death are intersecting in ever-more interesting ways. In this paper, I consider some emerging phenomena in online memorialisation and mourning practices and the specific features of online social networks that might licence the claim that the dead somehow live on through their online presence. I situate these practices within a phenomenology of grief that accounts for the ways in which the dead can persist as moral patients (a topic of considerable controversy in analytic metaphysics). We then consider divergent intuitions concerning the survival of others and our own survival through online media, and show how this case illuminates an important difference between persons and selves within contemporary philosophy of personal identity. Ultimately, the online persistence of the dead helps bring into view a deep ontological contradiction implicit in our dealings with death: the dead both live on as objects of duty and yet completely cease to exist.

. . . Why can everyone else survive their death but I can't? The answer is that persons can, in some dimensions of their identity, survive their deaths in ways that selves cannot. Selves are rooted in first-personal, present-tense experience in ways that are quite different to persons, and this allows persons to persist in circumstances where the associated self has dropped out of existence. So while from my perspective the people I've lost to death survive in some tragically reduced and ontologically ambiguous form, from their own perspective they do not survive their death at all—because their death is precisely the loss of the self that constitutes their own perspective. One can only survive one's death second- or third-personally, so from a first-personal perspective, no one else is any better off than me.

This is not to say that there are two objects with different persistence conditions, the self and the person, and that persons have wider diachronic extensions than selves. In an important sense, selves have no diachronic extension: they're always a feature of the present moment, albeit one that then appropriates the person's past or future. When I look at the Facebook profile of a dead friend, their person persists in the form of their extended phenomenality, in an essentially timeless way: they existed before they died and, in this much thinner but still phenomenally significant form, they'll continue to exist tomorrow. When I consider my own survival however, I find that I cannot identify in the necessary sense with the identity figured in my Facebook profile such that I could regard its continued existence as constituting the survival of the self I experience myself as being here and now.

The online space offers many opportunities to become someone else, to maintain anonymity or create a new identity unrelated to one's embodied existence. Online social networks, by contrast, represent a technology for articulating, expressing and expanding the agential and phenomenal reach of our anchored, socially recognised, intersubjective identity. These online identities do not, I think, represent a truly new form of identity, but they do extend our existing identities in compelling ways, and considerably enhance our posthumous phenomenality in the minds of others. So in an important sense, they genuinely do help the dead dwell among the living a little longer than they perhaps might have done. Yet they do so in a way that actually makes the separation between our first-personal extinction and our third-personal persistence all the more poignant.

4. A New Type of Immortality

In the 1970s, well before it was popular in academics to talk about death, psychologist Robert Kastenbaum (2008) suggested that our understanding of death is organized around how people think about, behave in light of, and structure their death experiences. Death systems, as he calls them, vary across cultural and historical periods but have common elements: people, places, times, objects, and symbols. These come together to construct what he calls "making sense of death." Recently, grief counselor Carla Sofka, Kathleen R. Gilbert, and Illene Noppe Cupit (2012) have looked at Kastenbaum's view of death in light of changes brought about by digital technology. What they find is that the elements in death systems (namely, people, places, times, objects, and symbols) change in the digital world, thereby challenging traditional understandings of death. For instance, a traditional view of death is tied to concrete or physical objects located in time and space. A digital understanding of death is less tied to actual physical objects in time and space and more associated with a less bounded sense of objects, time, and place (Sofka, Gilbert, and Cupit 2012, 8). This shift transforms our view and experience of death. Such reflections are among those of an emerging literature on death in the digital universe.

Reading: "Thanatechnology as a Conduit for Living, Dying, and Grieving in Contemporary Society" (Sofka, Cupit, and Gilbert 2012, 4–6). Reprinted with the permission of Springer Publishing Company, Inc.

It is a universal truth that at some point in each person's life, he or she will experience the death of a significant other. Although dealing with illness, death, and grief is common to all individuals, the emotional, cognitive, and physical reactions that accompany these experiences are now known to be affected by individual differences, culture, and historical period, causing each person's experience to be unique.

In many respects, grief reflects a social construction of reality referring to the meanings of death, dying, and grief that evolve from people who share time, place, and culture. Although individuals may develop their own unique viewpoint with respect to what death means, sharing that meaning with others helps to validate that perspective and consolidate their concepts about death. For example, in some cultures, the spirit of the deceased continues to be socially active and be a member of the community even after corporeal death. Contemporary societies that primarily adhere to a medical model posit that social interaction between the deceased and the living community ends with physical death. However, are these views about death and grief changing as a result of the current technological revolution?

Consider the following examples: In the not too distant past, dying individuals quietly "passed away," often hidden from view in a hospital bed. It is not uncommon . . . for terminally ill patients to describe their journey in intimate detail to family and friends—perhaps even millions of "strangers"—on a blog. As described in detail by Moore . . . , patients can interact with friends and family "in person" with the help of thanatechnology. Does an increased level of communication and "openness" about illness, dying, and death have an impact on dying persons and their dying? Does increased access to information lead to a greater understanding of the disease process? Do we seek different treatment options after we (and friends and family) "surf the net" for hopeful solutions? For the survivors, does it change how they cope with the death?

In the virtual world, if you have an online presence in life, you have the issue of your "digital legacy" to consider as you prepare for your eventual death. Hans-Peter Brondmo, the head of social software and services at Nokia in San Francisco, describes one's accumulation of websites, blogs, and personal profiles on various social networking sites (SNSs) and other online records of our existence as one's "digital soul."

After acknowledging that "we are the first people in history to create vast on-line records of our lives," Paul-Choudhury poses a question worth considering: "How much of it will endure when we are gone?" An individual can preserve the way he or she was prior to illness using images, video, or voiceprint. This can be done with the assistance of a tech-savvy loved one (e.g., see http://kathrynoates .org/, a website created to preserve her memory by her husband, Sumit Paul-Choudhury) or by purchasing the services of a company such as Deathswitch, Digital Legacy, or Virtual Eternity.

In the new world of thanatechnology, individuals want their "online soul" preserved to create a "technology heirloom" or even to continue to commu-nicate after death (posthumous messaging via e-mails; see www.letterfrombe yond.com/). A new cottage industry has evolved to help with the tasks of digital estate planning (e.g., www.digitalestateservices.com/), or you can "do it your-self." To learn more, consider consulting *Your Digital Afterlife* by Carroll and Romano, a book that answers the fundamental philosophical (and dare we say thanatechnological) question: "What happens to my digital stuff when I die?" In the contemporary digitized world, getting one's affairs in order may need to in-clude preparing a "digital will."

On the other hand, Mayer-Schonberger encourages us to consider the pos-sibility that "our tools for recording what we see, experience, and think have be-come so easy to use, inexpensive and effective that it is easier to let information accumulate in our 'digital external memories' than it is to bother deleting it. For-getting has become costly and difficult, while remembering is inexpensive and easy." Perhaps, as Walker notes, "a great deal of our digital expression is simple communication about the present, 'intentionally ephemeral.'" Should we bother to preserve it?

With the rising popularity of SNSs came the challenging problems faced by administrators of these sites when deciding upon a policy to handle the death of a member. Boddy aptly titled his article "Ghosts in the Machines" and de-scribed the reactions of family, friends, and even strangers to one deceased social media user's continued online presence as ranging from comfort to un-ease. Depending upon the policies in place on a particular site and the deci-sions made by family and friends, the digital legacy of an individual could be removed quickly following an individual's death or may remain indefinitely in cy-berspace. Facebook calls this process "memorializing" a deceased user's site to allow only confirmed friends to see the profile or locate it during a search. In addition to that continuing online presence, thanatechnology makes it possible for your digital presence to be experienced in other places affiliated with dying, death, and grief. The Internet Patrol describes the "Serenity Panel," a solar-powered video headstone that allows visitors to the cemetery to celebrate the life of the deceased. Japanese gravestones use two-dimensional bar codes (QR codes), which when scanned by a visitor's cellphone allows the visitor to view photos, videos, and other information about the deceased. Family members are able to view a log of who visits with a special device that keeps a record each time the code is scanned. VirtualEternity.com, a company whose motto is "forever made possible," helps a customer to convert the personal data that you provide into an avatar. Walker describes this "intellitar" as the company calls it "sort of like one of those chatbots that some online companies use for automated but more humanish customer service." It is designed to give users the gift of immortality.

With the profiles of the deceased remaining indefinitely on SNSs and sophisticated digital programming that can cause the deceased to once again "come alive at particular places and times through video images or voiceprint," is thanatechnology creating a type of digital immortality? The impact of these decisions and dilemmas regarding "digital ghosts" remain largely unstudied.

The availability of technology has inspired us to create ways to modulate the pain of loss through the use of computers and handheld devices. Mourning rituals now occur online, including virtual funerals and memorial services, Facebook tributes, and in online communities of bereavement. Although thanatechnology has clearly changed how, when, and where we grieve, has it changed our grief? The virtual world has truly opened up new issues and questions for the thanatologist. If the shared meanings of death, dying, and grief are culturally defined, would the globalization of technology ultimately create a uniform sense of the meaning of death and the ensuing grieving process? As de Vries and Moldaw argue, the World Wide Web has democratized the grief process, challenging who is entitled to grieve and how they should grieve. New online rituals may evolve within the global community, transcending distance, time, reality, diverse beliefs about death, and traditional expectations for the "proper" way to mourn.

C. REFLECTIONS: ONLINE IDENTITY AND OBITUARY

1.(*) Continue the discussions by Patrick Stokes and Carla Sofka and her colleagues. How does the use of digital technology alter our understanding of death? How does it alter how we understand what it means to be human? How does it change the people, places, times, objects, and symbols typically associated with death? Be sure to develop your thinking in light of the discussions and give details and examples.

2.(*) Given Richard Jones's question, and in light of Hans Moravec's contention that technological immortality is only a few generations away, address the following: Assuming that digital immortality is possible, would you choose it? If so, why? If not, why not? Develop your thoughts on the matter in light of the discussions and give reasons for the position you hold.

3.   This is a continuation of your advance planning in preparation for death:
   a. Write your obituary (or announcement of your death that appears in a local, regional, national, or international news outlet). You may wish to go online and read a few obituaries or instructions for writing obituaries to get a sense of how they are structured. As you did in the exercise on pricing out your end-of-life care (in chapter 6, C), you will need in this exercise to designate the age at which you die and write from that perspective.
   b. One of the lessons from Chris Faraone is that we need to make plans for what will happen to the digital information and sites we leave behind after we die. What would you like done with such data and sites? Would you like your digital information and sites to continue or to be erased or shut down? If you want them continued, why? If you want them erased or shut down, who will you ask to be responsible for this task? What steps do you need to take to ensure that your digital information and sites are handled as you wish after your death?

D. FURTHER READINGS

Barrow, John, and Frank Tipler. 1988. *The Anthropic Cosmological Principle.* New York: Oxford University Press.

Bell, Gordon, and Jim Gray. 2001. "Digital Immortality." *Communications of the ACM* 44 (3): 29–31. https://gordonbell.azurewebsites.net/CGB%20Files/CACM%20Digital%20Immortality%20with%20Gray%200103%20c.pdf.

Death and Digital Legacy website. www.deathanddigitallegacy.com/.

Digital Beyond website. www.thedigitalbeyond.com/.

Doka, Kenneth J., and Terry L. Martin. 2010. *Grieving beyond Gender: Understanding the Ways Men and Women Mourn.* New York: Taylor and Francis.

Faure, G. 2009. "How to Manage Your Online Life When You're Dead." *Time*, August 18. www
.time.com/time/business/article/0,8599,1916317,00.html.

Fletcher, D. 2009. "What Happens to Your Facebook after You Die?" *Time*, October 28. www
.time.com/time/business/article/0,8599,1932803,00.html.

Moravec, Hans. 1990. *Mind Children: The Future of Robots and Human Intelligence*. Cambridge, MA: Harvard University Press.

Chapter 8

An Existential Phenomenon of Life

A. CONTEXT

Many of the frameworks for death presented in previous chapters seek to provide a universal account that is applicable to all people. Consider, for instance, that proponents of reincarnation believe that their accounts of death cross historical and cultural barriers. We can say the same about accounts of death as resurrection, physical disintegration, psychological disintegration, medical immortality, and digital immortality. Yet there are those who reject the possibility of a universal account of death. Here, we find the existentialists and phenomenologists of the late modern and early contemporary Western periods of thought. Generally put, "existentialists" are those philosophers, from the nineteenth century to the present, who emphasize the meaning and choice of actual living beings (Kramer and Wu 1988, 241). "Phenomenologists" are those philosophers, from the nineteenth century to the present, who investigate human life as it appears to consciousness in terms of meaning, rather than how it is represented when logical thought patterns are imposed on it (Kramer and Wu 1988, 238). In both traditions, knowledge is constituted by the lived experience.

Although the readings from the existentialists and phenomenologists on death are some of the more challenging philosophical ones, they are uniquely insightful. Here is a simplified version of their message about death. Generally put, for existentialists and phenomenologists, death is a phenomenon *experienced* by humans. Given that we cannot experience death, we cannot know death itself. We cannot know death itself because we can know death only through the experience of the death of another, or through the death of some aspect of the self while we are living. It follows that we cannot have a universal metaphysics of death. We can only have an experience of death that is unique to each of us. Consequently death varies among us and leads to many particular "predicaments," such as knowing that we will die, yet not knowing about death itself, and having a sense of immortality or life after death, yet not being able to grasp the concept fully. What follows are selections on death from the notable existentialists Søren Aabye Kierkegaard, John-Paul Sartre, and Simone de Beauvoir as well as the acclaimed phenomenologist Martin Heidegger.

B. PERSPECTIVES

1. One's Hope and Despair

Danish philosopher Søren Aabye Kierkegaard (1813–55) was a profound and prolific writer in the Danish "golden age" of intellectual and artistic activity. His writings contribute significantly to the existentialist tradition, philosophy of religion, and Christian morality. Kierkegaard has notable writings on death. For Kierkegaard, death is decisive: To be human is to know that I will die. It is to be conscious of the fact that I will die, even though I can never fully experience death. I cannot fully experience death because that which allows experience (i.e., sensory perception) is gone. In reaction to this paradoxical state of knowing but not knowing, humans tend to despair or lose hope. As Kierkegaard says, "When death is the greatest danger, one hopes for life; but when one becomes acquainted with an even more dreadful danger, one hopes for death. So when the danger is so great that death has become one's hope, despair is the disconsolateness of not being able to die" ([1849] 1954, 151). As we learn in the following passage from *Sickness unto Death*, death indicates "the eternal in man" ([1849] 1954, 153). It indicates what is eternal in oneself because one cannot experience death and thus can neither rid oneself of life's burdens nor become nothing ([1849] 1954, 151). We are led to this: in the particularity of one's death, one finds one's eternal identity. In the end, one is not able to die. So death, as Kierkegaard tells us, consists "in not being able to die" ([1849] 1954, 154).

Reading: "Despair Is 'the Sickness unto Death,'" in *The Sickness unto Death* (Kierkegaard [1849] 1954, 150–54). Reprinted with the permission of Princeton University Press.

The concept of the sickness unto death must be understood, however, in a peculiar sense. Literally it means a sickness the end and outcome of which is death. Thus one speaks of a mortal sickness as synonymous with a sickness unto death. In this sense, despair cannot be called the sickness unto death. But in the Christian understanding of it death itself is a transition unto life. In view of this, there is from the Christian standpoint no earthly, bodily sickness unto death. For death is doubtless the last phase of the sickness, but death is not the last thing. If in the strictest sense we are to speak of a sickness unto death, it must be one in which the last thing is death, and death the last thing. And this precisely is despair.

Yet in another and still more definitive sense despair is the sickness unto death. It is indeed very far from being true that, literally understood, one dies of this sickness, or that this sickness ends with bodily death. On the contrary, the torment of despair is precisely this, not to be able to die. So it has much in common with the moribund when he lies and struggles with death, and cannot die. So to be sick *unto* death is, not to be able to die—yet not as though there were hope of life; no, the hopelessness in this case is that even the last hope, death, is not available. When death is the greatest danger, one hopes for life; but when one becomes acquainted with an even more dreadful danger, one hopes for death. So when the danger is so great that death has become one's hope, despair is the disconsolateness of not being able to die.

It is in this last sense that despair is the sickness unto death, this agonizing contradiction, this sickness in the self, everlasting to die, to die and yet not to die, to die the death. For dying means that it is all over, but dying the death means to live to experience death; and if for a single instant this experience is possible, it is tantamount to experiencing it forever. If one might die of despair as one dies of a sickness, then the eternal in him, the self, must be capable of dying in the same sense that the body dies of sickness. But this is an impossibility; the dying of despair transforms itself constantly into a living. The despairing man cannot die; no more than "the dagger can slay thoughts" can despair consume the eternal thing, the self, which is the ground for despair, whose worm dieth not, and whose fire is not quenched. Yet despair is precisely *self*-consuming, but it is an impotent self-consumption which is not able to do what it wills; and this impotence is a new form of self-consumption, in which again, however, the despairer is not able to do what he wills, namely, to consume himself. This is despair raised to a higher potency, or it is the law for the potentiation. This is the hot incitement, or the cold fire in despair, the gnawing canker whose movement is constantly inward, deeper and deeper, in impotent self-consumption. The fact that despair does not consume him is so far from being any comfort to the despairing man that it is precisely the opposite, this comfort is precisely the torment, it is precisely this that keeps the gnawing pain alive and keeps life in the pain. This precisely is the reason why he despairs—not to say despaired— because he cannot consume himself, cannot get rid of himself, cannot become nothing. This is the potentiated formula for despair, the rising of the fever in the sickness of the self.

So to despair over something is not yet properly despair. It is the beginning, or it is as when the physician says of a sickness that it has not yet declared itself. The next step is the declared despair, despair over oneself. A young girl is in despair over love, and so she despairs over her lover, because he died, or because he was unfaithful to her. This is not a declared despair; she is in despair over herself. This self of hers, which, if it had become "his" beloved, she would have been rid of in the most blissful way, or would have lost, this self is now a torment to her when it has to be a self without "him"; this self would have been to her her riches (though in another sense equally in despair) has now become to her a loathsome void, since "he" is dead, or it has become to her an abhorrence, since it reminds her of the fact that she was betrayed. Try it now, say to such a girl, "Though art becoming thyself," and thou shalt hear her reply, "Oh, no, the torment is precisely this, that I cannot do it."

To despair over oneself, in despair to will to be rid of oneself, is the formula for all despair, and hence the second form of despair (in despair at willing to be oneself) can be followed back to the first (in despair at not willing to be oneself), just as in the foregoing we resolved the first into the second (cf. I). A despairing man wants despairingly to be himself, he will not want to get rid of himself. Yet, so it seems; but if one inspects more closely, one perceives that after all the contradiction is the same. That self which he despairingly wills to be is a self which he is not (for to will to be that self which one truly is, is indeed the opposite of despair); what he really wills is to tear his self away from the Power which constituted it. But notwithstanding all his despair, that Power is the stronger, and it compels him to be the self he does not will to be. But for all that he wills to be rid of himself, to be rid of the self which he is, in order to be the self he himself has chanced to cho[o]se. To be the *self* as he wills to be would be his delight (though in another sense it would be equally in despair), but to be compelled to be the *self* as he does not will to be is his torment, namely, that he cannot get rid of himself.

Socrates proved the immortality of the soul from the fact that the sickness of the soul (sin) does not consume it as sickness of the body consumes the body. So also we can demonstrate the eternal in man from the fact that despair cannot consume his self, that this precisely is the torment of contradiction in despair. If there were nothing eternal in man, he could not despair; but if despair could consume his self, there would still be no despair.

Thus it is that despair, this sickness in the self, is the sickness unto death. The despairing man is mortally ill. In an entirely different sense than can appropriately be said of any disease, we may say that the sickness has attacked the noblest part; and yet the man cannot die. Death is the last phase of the sickness, but death is continually the last. To be delivered from this sickness by death is an impossibility, for the sickness and its torment . . . and death consist in not being able to die.

This is the situation in despair. And however thoroughly it eludes the attention of the despairer, and however thoroughly the despairer may succeed (as in the case of that kind of despair which is characterized by unawareness of being in despair) in losing himself entirely, and losing himself in such a way that it is not noticed in the least—eternity nevertheless will make it manifest that his situation was despair, and it will so nail him to himself that the torment nevertheless remains that he cannot get rid of himself, and it becomes manifest that he was deluded in thinking that he succeeded. And thus it is eternity must act, because to have a self, to be a self, is the greatest concession made to man, but at the same time, it is eternity's demand on him.

2. Possibility of the Absolute Impossibility of Da-sein

The German philosopher Martin Heidegger (1889–1976) has contributed to the diverse fields of phenomenology, existentialism, political theory, psychology, and theology. His critiques of traditional metaphysics have been embraced by leading theorists of postmodernism, those who reject the view that there is a so-called universal metanarrative that encompasses all narratives and provides a uniform foundation or explanatory framework for knowledge claims (Lyotard 1984). In keeping with the theme that a metanarrative on death is unavailable, Heidegger defends the view that death is a "phenomenon of life" ([1927] 1996, 229). It is a lived experience of life and that toward which human life aims. It is a lived experience because I can know only through experience, which ends at death. It is the absolute possibility or potentiality of nonbeing. As he says in the following passage from *Being and Time*, death is the "possibility of the absolute impossibility of Da-sein" ([1927] 1996, 232). Here, *Da-sein* refers to "being-in-the-world," a being who is neither a subject nor an object but one *engaged* in the world. On this interpretation, there can be no "abstract being"; there is only a being who emerges out of his or her local experience and engagements. In other words, there is no abstract "I"; there is only an "I" that engages with the world that is "my" world and experiences it in particular and unique ways. Similarly, there can be no "abstract death"; there is only the death of an actual being who is engaged in his or her world. And the death of an actual being marks the possibility of nonexistence of a living being and the impossibility of being-in-the-world. It marks a threshold of experience and existence of a living being. It indicates "the possibility of no-longer-being-able-to-be-there" ([1927] 1996, 232).

Reading: *Being and Time* (Heidegger [1927] 1996, 229–31, 232 [§§49 and 50]). Reprinted with the permission of Walter de Gruyter GmbH.

49. *How the Existential Analysis of Death Differs from Other Possible Interpretations of This Phenomenon*

The unequivocal character of the ontological interpretation of death should be made more secure by explicitly bringing to mind what this interpretation can *not* ask about and where it would be useless to expect instructions.

In the broadest sense, death is a phenomenon of life. Life must be understood as a kind of being to which belongs a being-in-the-world. It can only be defined in a privative orientation to Da-sein. Da-sein, too, can be considered as pure life. For the biological and physiological line of questioning, it then moves into the sphere of being which we know as the world of animals and plants. In this field, dates and statistics about the life-span of plants, animals and men can be ontologically ascertained. Connections between the life-span, reproduction and growth can be known. The "kinds" of death, the causes, "arrangements" and ways of its occurrence can be investigated.

An ontological problematic underlies this biological and ontic investigation of death. We must still ask how the essence of death is defined in terms of the essence of life. More or less clarified preconceptions of life and death are operative in it. These preliminary concepts need to be sketched out in the ontology of Da-sein. Within the ontology of Da-sein, which has priority over an ontology of life, the existential analytics of death is *subordinate* to the fundamental constitution of Da-sein. We called the ending of what is alive *perishing*. Da-sein, too, "has" its physiological death of the kind appropriate to anything that lives and has it not ontically in isolation, but as also determined by its primordial kind of being. Da-sein too can end without authentically dying, though on the other hand, qua Da-sein, it does not simply perish. We call this intermediate phenomenon its *demise*. Let the term *dying* stand for the *way of being* in which Da-sein is *toward* its death. Thus we can say that Da-sein never perishes. Da-sein can only demise as long as it dies. The medical and biological inquiry into *demising* can attain results which can also become significant ontologically if the fundamental orientation is ensured for an existential interpretation of death. Or must sickness and death in general—even from a medical point of view, too—be conceived primarily as existential phenomena?

The existential interpretation of death is prior to any biology and ontology of life. But it is also the foundation for any biographical-historical or ethnologico-psychological inquiry into death. A "typology" of "dying" characterizing the states and ways in which a demise is "experienced" already presupposes the concept of death. Moreover, a psychology of "dying" rather gives information about the "life" of the "dying person" than about dying itself. That is only a reflection of the fact that when Da-sein dies—and even when it dies authentically—it does not have to do so with an experience of its factical demise, or in such an experience. Similarly, the interpretations of death in primitive peoples, of their behavior toward death in magic and cult, throw light primarily on the understanding of Da-sein; but the interpretation of this understanding already requires an existential analytic and a corresponding concept of death.

The ontological analysis of being toward the end, on the other hand, does not anticipate any existential stand toward death. If death is defined as the "end" of Da-sein, i.e., of being-in-the-world, no ontic decision has been made as to whether "after death" another being is still possible, either higher or lower, whether Da-sein "lives on" or even, "outliving itself," is "immortal." Nor is anything decided ontically about the "otherworldly," as if norms and rules for behavior toward death should be proposed for "edification." But our analysis of death remains purely "this-worldly" in that it interprets the phenomenon solely with respect to the question of how it *enters into* actual Da-sein as its possibility of being. We cannot even *ask* with any methodological assurance about what "is after death" until death is understood in its full ontological essence. Whether such a question presents a possible *theoretical* question at all is not to be decided here. The this-worldly, ontological interpretation of death comes before any ontic, other-worldly speculation.

Finally, what people would like to discuss under the rubric of a "metaphysics of death" lies outside the scope of an existential analysis of death. The questions of how and when death "came into the world," what "meaning" it can and should have as an evil and suffering in the whole of beings—these are questions that necessarily presuppose an understanding not only of the character of being of death, but the ontology of the whole of beings as a whole and the ontological clarification of evil and negativity in particular.

The existential analysis is methodologically prior to the questions of a bi-ology, theodicy, and theology of death. Taken ontically, the results of the analysis show the peculiar *formality* and emptiness of any ontological characterization. However, that must not make us blind to the rich and complex structure of the phenomenon. Since Da-sein never becomes accessible at all as something ob-jectively present, because being possible belongs, in its own way, to its kind of being, even less may we expect to simply read off the ontological structure of death, if indeed death is an eminent possibility of Da-sein.

On the other hand, our analysis cannot be supported by an idea of death that has been devised arbitrarily and at random. We can restrain this arbitrari-ness only by giving beforehand an ontological characterization of the kind of being in which the "end" enters into the average everydayness of Da-sein. For this we need to envisage fully the structures of everydayness worked out earlier. The fact that existential possibilities of being toward death have their resonance in an existential analysis of death is implied by the essence of the ontological in-quiry. All the more explicitly, then, must an existential neutrality go together with the existential conceptual definition, especially with regard to death, where the character of Da-sein can be revealed most clearly of all. The existential problem-atic aims solely at developing the ontological structure of the being *toward* the end of Da-sein. . . .

50. *A Preliminary Sketch of the Existential and Ontological Structure of Death*

. . . Death is a possibility of being that Da-sein always has to take upon itself. With death, Da-sein stands before itself in its ownmost potentiality of being. In this possibility, Da-sein is concerned about its being-in-the-world absolutely. Its death is the possibility of no-longer-being-able-to-be-there. When Da-sein is imminent to itself as this possibility, it is *completely* thrown back upon its ownmost poten-tiality of being. Thus imminent to itself, all relations to other Da-sein are dissolved in it. This non-relational ownmost possibility is at the same time the most extreme one. As a potentiality of being, Da-sein is unable to bypass the possibility of death. Death is the possibility of the absolute impossibility of Da-sein. Thus *death* reveals itself as the ownmost *non-relational possibility not to be bypassed*. As such, it is *an eminent* imminence. Its existential possibility is grounded in the fact that Da-sein is essentially disclosed to itself, in the way of being-ahead-of-it-self. This structural factor of care has its most primordial concretion in being-to-ward-death. Being-toward-the-end becomes phenomenally clearer as being to-ward the eminent possibility of Da-sein which we have characterized.

3. A Boundary

French philosopher Jean-Paul Sartre (1905–80) is a well-known existentialist of the twentieth century who contributed to philosophical, literary, and political discussions. In 1964, Sartre was awarded a Nobel Peace Prize for Literature but declined to accept it because he did not want to become a public institution representing something he did not wish to support. Like Kierkegaard and Heidegger, Sartre had a notable interest in death. For Sartre, death represents a major change in life, a threshold that one can conceive in terms of one's life. As he says in *Being and Nothingness*, "Death has always been—rightly or wrongly is what we can not yet determine— considered as a final boundary of human life.... Death is a human phenomenon; it is the final phenomenon of life and is still life. As such it influences the entire life by a reverse flow. Life is limited by life" ([1943] 1956, 532). In other words, death is the last chapter of one's own book. What we can know about death depends on life experience. Further, this knowledge is *my* experience. As Sartre says, "Death thus recovered does not remain simply human; it becomes mine" ([1943] 1956, 532). Given this, it will be impossible to arrive at a universal metanarrative of death because any account of death will be a reflection of the experience of an individual human being based on his or her experience with death. In addition, for Sartre, death is "absurd" ([1943] 1956, 547). As he puts it, "It is absurd that we are born; it is absurd that we die" ([1943] 1956, 547). It is absurd because death makes no logical sense. There is nothing in life that logically requires its end. There is nothing in death that is logically required to be a certain way. In what follows, Sartre offers the view that an understanding of death is *my* experience of death.

Reading: "My Death," in *Being and Nothingness* (Sartre [1943] 1956, 531–32, 543, 547–48). Reprinted with the permission of Philosophical Library, Inc.

After death had appeared to us as pre-eminently non-human since it was what there was on the other side of the "wall," we decided suddenly to consider it from a wholly different point of view—that is, as an event of human life. This change is easily explained: death is a boundary, and every boundary (whether it be final or initial) is a *Janus bifrons*. Whether it is thought of as adhering to nothingness of being which limits the process considered or whether on the contrary it is revealed as adhesive to the series which it terminates, in either case it is a being which belongs to an existent process and which in a certain way constitutes the meaning of the process. Thus the final chord of a melody always looks on the one side toward silence—that is, toward the nothingness of sound which will follow the melody; in one sense it is made with the silence since the silence which will follow is already present in the resolved chord as its meaning. But on the other side it adheres to the *plenum* of being which is the melody intended; without the chord this melody would remain in the air, and this final indecision would flow back from note to note to confer on each of them the quality of being unfinished.

Death has always been—rightly or wrongly is what we can not yet determine—considered as a final boundary of human life. . . . Death is a human phenomenon; it is the final phenomenon of life and is still life. As such it influences the entire life by a reverse flow. Life is limited by life; it becomes like the world of Einstein, "finite but unlimited." Death becomes the meaning of life as the resolved chord is the meaning of the melody. There is nothing miraculous in this; it is the one term in a series under consideration, and as one knows, each term of a series is always present in all the terms of the series.

But death thus recovered does not remain simply human; it becomes *mine*. By being interiorized it is individualized. Death is no longer the great unknowable which limits the human; it is the phenomenon of *my* personal life which makes of this life a unique life—that is, a life which does not begin again, a life in which one never recovers his stroke. Hence I become responsible for *my* death as for my life. Not for the empirical and contingent phenomenon of my decease but for this character of finitude which causes my life like my death to be *my* life. . . .

Thus the very existence of *death* alienates us wholly in our own life to the advantage of the Other. To be dead is to be a prey for the living. This means therefore that the one who tries to grasp the meaning of his future death must discover himself as the future prey of others. We have here therefore a case of alienation. . . .

. . . It is absurd that we are born; it is absurd that we die. On the other hand, this absurdity is presented as the permanent alienation of my being-possibility which is no longer *my* possibility but that of the Other. It is therefore an external and factual limit of my subjectivity! . . .

Mortal represents the present being which I am for the Other; *dead* represents the future meaning of my actual for-itself for the Other. . . .

. . . Death is not an obstacle to my projects; it is only a destiny of these projects elsewhere. And this is not because death does not limit my freedom but because freedom never encounters this limit. I am not "free to die," but I am a free mortal. Since death escapes my projects because it is unrealizable, I myself escape death in my very project. Since death is always beyond my subjectivity, there is no place for it in my subjectivity. This subjectivity does not affirm itself *against* death but independently of it although this affirmation is immediately alienated. Therefore we can neither think of death nor wait for it nor arm ourselves against it; but also our projects as projects are independent of death—not because of our blindness, as the Christians say, but on principle. And although there are innumerable possible attitudes with which we may confront this unrealizable which "in the bargain" is to be realized, there is no place for classifying these attitudes as authentic or unauthentic since we always die *in the bargain*.

4. Very Easy

Simone de Beauvoir (1908–86) is a well-known existentialist of the twentieth century who contributed to philosophical, literary, and political discussions. She authored numerous works on feminism, politics, and ethics. Her book *The Second Sex* ([1949] 1989) is considered a revolutionary feminist treatise on women's oppression and liberation that appeared well before numerous other accounts in the literature. Beauvoir is also known for being poignantly honest in her writing. In *A Very Easy Death*, Beauvoir shares with us her experience of witnessing her mother's, or Maman's, death from cancer. While she claims that her Maman experiences "a very easy death" (1985, 95), the death is not easy. Her Maman's death is fraught with competing emotions, bodily responses, and terrifying insights. As Beauvoir says, "There is no such thing as a natural death: nothing that happens to a man is ever natural, since his presence calls the world into question" (1985, 106). Consider these passages from Beauvoir's *A Very Easy Death*.

Reading: *A Very Easy Death* (Beauvoir 1985, 62, 81, 94–95, 106)

I did not particularly want to see Maman again before her death; but I could not bear the idea that she should not see me again. Why attribute such importance to a moment since there would be no memory? There would not be any atonement either. For myself I understood, to the innermost fibre of my being, that the absolute could be enclosed within the last moments of a dying person.

Then suddenly she cried out, a burning pain in her left buttock. It was not at all surprising. Her flayed body was bathing in uric acid that oozed from her skin. . . . All tense on the edge of shrieking, she moaned, "It burns, it's awful; I can't stand it. I can't bear it any longer." And half sobbing, "I'm so utterly miserable," in that child's voice that pierced me to the heart. How completely alone she was! I touched her, I talked to her; but it was impossible to enter into her suffering. . . . Nothing on earth could possibly justify these moments of pointless torment.

She suddenly cried, "I can't breathe!" Her mouth opened, her eyes stared wide, huge in that wasted, ravaged face: with a spasm she entered into a coma. . . .

"But Madame," replied the nurse, "I assure you it was an easy death."

For indeed, comparatively speaking, her death was an easy one. . . . I thought of all those who have no one to make that appeal to: what agony it must be to feel oneself a defenceless thing, utterly at the mercy of indifferent doctors and over-worked nurses. No hand on the forehead when terror seizes them; no sedative as soon as pain begins to tear them. . . . She [Maman] had a very easy death; an upper-class death.

There is no such thing as a natural death: nothing that happens to a man is ever natural, since his presence calls the world into question.

C. REFLECTIONS: BUCKET LIST

1.   The prior selections by Søren Kierkegaard, Martin Heidegger, Jean-Paul Sartre, and Simone de Beauvoir are often seen to be some of the more challenging readings on death. Yet at the same time they are often seen to be the most insightful. Draw from these readings as well as examples from your own life experience to make sense of what these authors tell us about death.

   a.(*) Put Kierkegaard's view of death into your own words. In what ways might your death be your hope? How might your death give "energy to life as nothing else does"? Draw from Kierkegaard's reading as well as examples from your own life experience to make sense of what Kierkegaard tells us.

   b.(*) Put Heidegger's view of death into your own words. How does a "metaphysics of death" lie "outside the scope of an existential analysis of death" in your life? How is death the "possibility of the absolute impossibility of Da-sein" in your life? Draw from Heidegger's reading as well as examples from your own life experience of death to make sense of what Heidegger tells us.

   c.(*) Put Sartre's view of death into your own words. How is your death a "boundary" in your life? How is death "mine"? How is it "absurd"? How does your death represent the future meaning of your "actual for-itself for the Other"? Draw from Sartre's reading as well as examples from your own life experience to make sense of what Sartre tells us.

d.(*) Put Beauvoir's view of death into your own words. How is Maman's death "easy"? How is Maman's death an "upper-class death"? Yet how is Maman's death so agonizing and terrorizing? Draw from Beauvoir's reading as well as examples from your own life experience to make sense of what Beauvoir tells us.

2.   This is a continuation of your advance planning in preparation for death. The 2007 movie *The Bucket List* (Warner Brothers Entertainment) made popular the exercise of writing one's own bucket list, an itemization of activities one wishes to complete before one dies. Such activities can include goals one wishes to accomplish, places one wishes to travel, skills one wishes to acquire, and so forth. Prepare your own bucket list. Reflect briefly on why you select the items you put on the list.

D. FURTHER READINGS

Barnes, Hazel E. 1959. *The Literature of Possibility: A Study in Humanistic Existentialism.* Lincoln: University of Nebraska Press.

Beauvoir, Simone de. 1985. *A Very Easy Death.* New York: Pantheon.

Buben, Adam, and Patrick Stokes. 2011. *Kierkegaard and Death.* Bloomington: Indiana University Press.

Feifel, Herman. 1959. *The Meaning of Death.* New York: McGraw-Hill.

Kaufman, Walter. 1989. *Existentialism from Dostoyevsky to Sartre.* New York: Meridian.

Lyotard, Francois. 1984. *The Postmodern Condition: A Report on Knowledge.* Translated by G. Bennington and B. Massumi. Minneapolis: University of Minnesota Press.

Willis, Claire. 2014. *Lasting Words: A Guide to Finding Meaning toward the Close of Life.* Brattelboro, VT: Green Writers Press.

Part II

THE VALUE OF DEATH

In part II, "The Value of Death," death is understood in terms of a variety of evaluations. The question in this section of the analysis is "How do I or we value death?" In philosophy and religious studies, the study of values is called axiology (from the Greek *axiā*, meaning "value or worth," and *-ology*, meaning "study of"). Axiology is rooted in ancient quests to understand what and how we value. A *value* is a sign of significance and may be understood in terms of nonmoral and moral evaluations. *Nonmoral values* concern how we measure or assess persons, objects, or events. Examples of nonmoral values include aesthetic, instrumental, and economic assessments, among others. In discussions of death, for instance, death can be seen to be "bad" because it can be aesthetically horrific given a decomposing body. Death can be seen to be instrumentally detrimental to promoting the continuation of life. Death can be seen to be expensive, especially in cultures in which there is a market for death practices and services.

Moral or ethical values concern that which is praiseworthy and blameworthy, or right or wrong. They give voice to the expectations of others as well as the self on matters concerning what we or I ought to do or be. Examples of ethical values or principles include respect for a person (i.e., autonomy), the minimization of harm (i.e., nonmaleficence), the promotion of the best interest of another (i.e., beneficence), and the distribution of resources in a fair way (i.e., justice) (Beauchamp and Childress 2013). In discussions of death, for instance, death can be seen to be "praiseworthy" when someone risks his or her life in order to minimize harm or promote the welfare of others. Alternatively, death can be seen to be "blameworthy" when it is brought about by means that violate the autonomy of a person or that harm the self or another. Readers see some appeal to these concerns in this chapter, and more in part III to come.

The readings in this second part of our investigation of death explore a range of evaluative views of death. These include death as bad or good, to be feared or not, and to be grieved and how. Here evaluative views of death are presented through the lens of the nature of death, thus linking popular clinical and psychological topics on death with philosophical or sacred text discussions of the nature of death. In reading the following selections, look for ways in which we evaluate death and what this says about what it means to be human. Readers may find that they agree with some of the authors' claims and disagree with others. They are encouraged to note these points of agreement and disagreement with authors in preparation for the exercises that appear at the end of each chapter.

— Chapter 9 —

Bad or Good

A. CONTEXT

Death is often claimed to be bad, good, or both bad and good. For many, death is bad insofar as it hinders the interests or welfare of a person. Death can be considered bad or evil when it brings about a loss of that which is considered a good, such as a relationship with a loved one or the achievement of a future possibility. For others, death is good insofar as it promotes the interests of a person. For instance, death can be considered good when it brings freedom from pain and suffering in a context in which little can be done to alleviate such pain or suffering. For many, however, death is both bad and good, as in the case of a death of a loved one who has passed and is deeply missed but who is finally at rest after a long bout of illness. Whether we judge death to be bad or good turns on a host of considerations regarding how we judge death and the extent to which it undermines or protects autonomy, deters or enhances the interests of a person, and is in line with sacred text teachings and values.

In the previous chapters, we have already seen examples of those who view death as bad or good. As Daniel 12:2 says, with death comes either the bad of "reproaches and everlasting abhorrence" or the good of "everlasting life," depending on how one lives one's life and how it is judged by the Creator. Christ's death and resurrection, as reported in Matthew 27 and 28, provide pivotal hope for what will come to those who follow Christ. The alternative for those who do not follow Christ is separation from God, which Augustine considered as "the second death" (Augustine 2000, 412–14). As Surah 3:185 and 191 say, with death comes "Paradise" and everlasting life with the divine, as opposed to "Fire." For Sartre, since death is the phenomenon of *my* personal life that makes my life a unique life, death can be bad or good depending on how I experience it. The following selections illustrate additional accounts of death as judged to be bad or good. Selections from an African tale, the Buddhist *Dhammapada*, the Japanese *Kojiki*, and American philosopher Thomas Nagel reflect on the "badness" of death. Greek philosopher Plato, Chinese philosopher Chuang Tzu, and Roman emperor Marcus Aurelius consider the "goodness" of death.

B. PERSPECTIVES

1. Bad

a. A Perverted Message

In the African oral tradition, a given story has many variants and a common story reproduces the tale in one of many possibilities (Feldmann 1963, 13). Common in African folklore is a highly humanized animal trickster, such as a hare, who is used to convey a moral message. The most widespread myth concerning the origin of death is the tale of the "perverted message" (Feldmann 1963, 27). In a rendition found in tribal traditions of East Africa, God sends an insect to humans with the message that they will have eternal life. A hare hears the message, distorts it, and conveys what is called a "perverted" message to humans: "As I die and dying perish, in the same manner you also shall die and come wholly to an end" (Feldmann 1963, 107). The perverted message is that death has entered the human world and all shall perish. Humans come to believe what the hare says, so death becomes a reality for humans.

Reading: "The Perverted Message," in *African Myths and Tales* (Feldmann 1963, 107)

The Moon, it is said, once sent an insect to men, saying, "Go to men and tell them, 'As I die, and dying live, so you shall also die, and dying live.'"

The insect started with the message but while on his way was overtaken by the hare, who asked, "On what errand are you bound?" The insect answered, "I am sent by the Moon to men, to tell them that as she dies and dying lives, so shall they also die and dying live."

The hare said, "As you are an awkward runner, let me go." With these words he ran off, and when he reached men, he said, "I am sent by the Moon to tell you, 'As I die and dying perish, in the same manner you also shall die and come wholly to an end.'"

The hare then returned to the Moon and told her what he had said to men. The Moon reproached him angrily, saying. "Do you dare tell the people a thing which I have not said?"

With these words the Moon took up a piece of wood and struck the hare on the nose. Since that day the hare's nose has been slit, but men believe what Hare had told them.

b. A State of Woe

As previously discussed (see chapter 4, B.1.c), the Buddha (ca. 560–480 BCE) has left a significant tradition of teaching on life and death. Central to Buddha's teachings are what are called the "Four Noble Truths," which are found in his first sermon after his enlightenment and handed down to his followers. Briefly put, the "Four Noble Truths" are (1) to exist is to suffer, (2) self-centeredness is the cause of human suffering, (3) the cause of suffering can be understood and rooted out, and (4) suffering can be alleviated by following the Eightfold Path. The Eightfold Path includes (1) right understanding, (2) right purpose, (3) right speech, (4) right conduct, (5) right livelihood, (6) right effort, (7) right mindfulness, and (8) right meditation (Soccio 2004, 41–42). Central to the Buddha's teaching, "the idea of self is an imaginary, false belief which has no corresponding reality" and provides "harmful thoughts of 'me' and 'mine'" (Rahula 1974, 51).

There are a number of schools in Buddhism. The two main traditions in Buddhism are Mahayana and Theravada. The Mahayana tradition is the larger tradition and is known for its philosophy of yoga practice. The earliest scriptures attributed to the Buddha in the Mahayana tradition of northern Asia are known as the *Tripitaka* (meaning "three baskets") and include the *Vinaya Pitaka, Sutta Pitaka,* and *Abhidhamma Pitaka.* In the *Dhammapada,* which is the second book of the *Khuddaka Nikaya* in the *Sutta Pitaka,* the following is said about what happens at death to a man who does evil: "There are many evil characters and uncontrolled men wearing the saffron [i.e., monastic] robe. These wicked men will be born in states of woe because of their evil deeds" ("Nirayavagga" 2010 [*Dhammapada* §307]). Alternatively, the following is said about what happens at death to someone who does good: "Those who discern the wrong as wrong and the right as right—upholding right views, they go to realms of bliss" ("Nirayavagga" 2010 [*Dhammapada* §319]). In what follows, chapter 22 of the *Dhammapada* provides examples of those who are eligible to obtain a state of woe, *nirayavagga*, or hell.

Reading: "Nirayavagga: Hell," in the *Dhammapada* (2010 [22.306–19]). Reprinted with the permission of the Buddhist Publication Society.

306. The liar goes to the state of woe; also he who, having done (wrong), says, "I did not do it." Men of base actions both, on departing they share the same destiny in the other world.

307. There are many evil characters and uncontrolled men wearing the saffron robe. These wicked men will be born in states of woe because of their evil deeds.

308. It would be better to swallow a red-hot iron ball, blazing like fire, than as an immoral and uncontrolled monk to eat the alms of the people.

309. Four misfortunes befall the reckless man who consorts with another's wife: acquisition of demerit, disturbed sleep, ill-repute, and (rebirth in) states of woe.

310. Such a man acquires demerit and an unhappy birth in the future. Brief is the pleasure of the frightened man and woman, and the king imposes heavy punishment. Hence, let no man consort with another's wife.

311. Just as *kusa* grass wrongly handled cuts the hand, even so, a recluse's life wrongly lived drags one to states of woe.

312. Any loose act, any corrupt observance, any life of questionable celibacy — none of these bear much fruit.

313. If anything is to be done, let one do it with sustained vigor. A lax monastic life stirs up the dust of passions all the more.

314. An evil deed is better left undone, for such a deed torments one afterwards. But a good deed is better done, doing which one repents not later.

315. Just as a border city is closely guarded both within and without, even so, guard yourself. Do not let slip this opportunity (for spiritual growth). For those who let slip this opportunity grieve indeed when consigned to hell.

316. Those who are ashamed of what they should not be ashamed of, and are not ashamed of what they should be ashamed of—upholding false views, they go to states of woe.

317. Those who see something to fear where there is nothing to fear, and see nothing to fear where there is something to fear—upholding false views, they go to states of woe.

318. Those who imagine evil where there is none, and do not see evil where it is—upholding false views, they go to states of woe.

319. Those who discern the wrong as wrong and the right as right—upholding right views, they go to realms of bliss.

c. Pollution

Shintoism is the indigenous spirituality of the Japanese people. The term *Shinto* means "Way of the Gods." While it was coined in the nineteenth century, it refers to a long history of practices recorded in ancient texts. One of these texts is the *Kojiki*, which was written in the eighth century CE and captures the historical tradition of Japan. The *Kojiki* tells the story of Izanami (His Augustness the Male-Who-Invites) and Izanagi (Her Augustness the Female-Who-Invites), the two *kamis* or gods who give birth to offspring, which are the islands of Japan ([1919] 2012, 712). But then tragedy strikes. Izanagi dies after giving birth to the *kami* of fire, and she goes to a place called the Land of Darkness or Hades (known as *Yomi no Kuni*). Izanami misses her so much that he follows her and finds her there, only to be shocked by her advanced state of decay. He flees Hades and stops at a river to cleanse himself on his way back to the land of the living. As the *Kojiki* says, "Therefore the Great Deity the Male-Who-Invites said: 'Nay! hideous! I have come to a hideous and polluted land,—I have! So I will perform the purification of my august person.' So he went out to a plain [covered with] *ahagi* [bushclover] at a small river-mouth near Tachibana in Himuka in [the island of] Tsukushi, and purified and cleansed himself" ([1919] 2012, 44). The message here is that we best cleanse or detach ourselves from life's possessions and wants. What follows highlights a Shinto understanding of death taken from sections 9 and 10 of the *Kojiki*.

Reading: "The Land of Hades" and "The Purification of the August Person," in the *Kojiki* ([1919] 2012, 38–41, 44 [§§9 and 10])

[§9—The Land of Hades]
Thereupon [His Augustness the Male-Who-Invites], wishing to meet and see his younger sister Her Augustness the Female-Who-Invites, followed after her to the

Land of Hades [the world beneath the grave]. So when from the palace she raised the door and came out to meet him, His Augustness the Male-Who-Invites spoke, saying: "Thine Augustness my lovely younger sister! the lands that I and thou made are not yet finished making; so come back!" Then Her Augustness the Female-Who-Invites answered, saying: "Lamentable indeed that thou camest not sooner! I have eaten of the furnace of Hades. Nevertheless, as I reverence the entry here of Thine Augustness my lovely elder brother, I wish to return. Moreover I will discuss it particularly with the Deities of Hades. Look not at me!" Having thus spoken, she went back inside the palace; and as she tarried there very long, he could not wait. So having taken and broken off one of the end-teeth of the multitudinous and close-toothed comb stuck in the august left bunch [of his hair], he lit one light and went in and looked. Maggots were swarming and [she was] rotting, and in her head dwelt the Great-Thunder, in her breast dwelt the Fire-Thunder, in her left hand dwelt the Young-Thunder, in her right hand dwelt the Earth-Thunder, in her left foot dwelt the Rumbling-Thunder, in her right foot dwelt the Couchant-Thunder:—altogether eight Thunder-Deities had been born and dwelt there. Hereupon His Augustness the Male-Who-Invites, overawed at the sight, fled back, whereupon his younger sister Her Augustness the Female-Who-Invites said: "Thou hast put me to shame," and at once sent the Ugly-Female-of-Hades to pursue him. So His Augustness the Male-Who-Invites took his black august head-dress and cast it down, and it instantly turned into grapes. While she picked them up and ate them, he fled on; but as she still pursued him, he took and broke the multitudinous and close-toothed comb in the right bunch [of his hair] and cast it down, and it instantly turned into bamboo-sprouts. While she pulled them up and ate them, he fled on. Again later [his younger sister] sent the eight Thunder-Deities with a thousand and five hundred warriors of Hades to pursue him. So he, drawing the ten-grasp sabre that was augustly girded on him, fled forward brandishing it in his back hand; and as they still pursued, he took, on reaching the base of the Even-Pass-of-Hades, three peaches that were growing at its base, and waited and smote [his pursuers therewith], so that they all fled back. Then His Augustness the Male-Who-Invites announced to the peaches: "Like as ye have helped me, so must ye help all living people in the Central Land of Reed-Plains when they shall fall into troublous circumstances and be harassed!"—and he gave [to the peaches] the designation of Their Augustnesses Great-Divine-Fruit. Last of all his younger sister Her Augustness the Princess-Who-Invites came out herself in pursuit. So he drew a thousand-draught rock, and [with it] blocked up the Even-Pass-of-Hades, and placed the rock in the middle; and they stood opposite to one another and exchanged leave-takings; and Her Augustness the Female-

Who-Invites said: "My lovely elder brother, thine Augustness! If thou do like this, I will in one day strangle to death a thousand of the folks of thy land." Then His Augustness the Male-Who-Invites replied: "My lovely younger sister, Thine Augustness! If *thou* do this, *I* will in one day set up a thousand and five hundred parturition-houses. In this manner each day a thousand people would surely be born." So Her Augustness the Female-Who-Invites is called the Great-Deity-of-Hades. Again it is said that, owing to her having pursued and reached [her elder brother], she is called the Road-Reaching-Great-Deity. Again the rock with which he blocked up the Pass of Hades is called the Great-Deity-of-the-Road-Turning-Back, and again it is called the Blocking-Great-Deity-of-the-Door-of-Hades. So what was called the Even-Pass-of-Hades is now called the Ifuya-Pass in the Land of Idzumo.

[§10—The Purification of the August Person]
Therefore the Great Deity the Male-Who-Invites said: "Nay! hideous! I have come to a hideous and polluted land,—I have! So I will perform the purification of my august person." So he went out to a plain [covered with] *ahagi* at a small river-mouth near Tachibana in Himuka in [the island of] Tsukushi, and purified and cleansed himself. So the name of the Deity that was born from the august staff which he threw down was the Deity Thrust-Erect-Come-Not-Place. The name of the Deity that was born from the august girdle which he next threw down was the Deity Road-Long-Space.

d. Deprivation or Loss

Thomas Nagel (b. 1937) is an American philosopher and currently a professor of philosophy and law at New York University in New York City. His main areas of scholarly interest are philosophy of mind, political philosophy, and ethics. In the late 1970s, Nagel published an essay that gained wide popularity for how it addressed the "badness" of death and challenged the view that death is to be accepted as a natural part of life that is good or neutral, as advanced by Plato (see B.2.a in this chapter) and Epicurus (see chapter 10, B.2.b), respectively. For Nagel, death is bad not because of something about the event itself but because it deprives us of something that we desire. As he says, "If we are to make sense of the view that to die is bad, it must be on the ground that life is a good and death is a corresponding deprivation or loss, bad not because of any positive features but because of the desirability of what it removes" (1979, 4). In what follows, Nagel argues that death is bad because death involves the loss of future possibilities in an individual's life.

Reading: "Death," in *Mortal Questions* (Nagel 1979, 1, 4–5). Reprinted with the permission of Cambridge University Press.

If death is the unequivocal and permanent end of our existence, the question arises whether it is a bad thing to die.

There is a conspicuous disagreement about the matter: some people think death is dreadful; others have no objection to death *per se,* though they hope their own will be neither premature nor painful. Those in the former category tend to think those in the latter are blind to the obvious, while the latter suppose the former to be prey to some sort of confusion. On the other hand it can be said that life is all we have and the loss of it is the greatest loss we can sustain. On the other hand it may be objected that death deprives this supposed loss of its subject, and that if we realize that death is not an unimaginable condition of the persisting person, but a mere blank, we will see that it can have no value whatever, positive or negative. . . .

If we are to make sense of the view that to die is bad, it must be on the ground that life is a good and death is a corresponding deprivation or loss, bad not because of any positive features but because of the desirability of what it removes. We must now turn to the serious difficulties which this hypothesis raises, difficulties about loss and privation in general, and about death in particular.

Essentially, there are three types of problems. First, doubt may be raised whether *anything* can be bad for a man without being positively unpleasant to him: specifically, it may be doubted that there are any evils which consist merely in the deprivation or absence of possible goods, and which do not depend on someone's *minding* that deprivation. Second, there are special difficulties, in the case of death, about how the supposed misfortune is to be assigned to a subject at all. There is doubt both as to *who* its subject is, and as to *when* he undergoes it. So long as a person exists, he has not yet died, and once he has died, he no longer exists; so there seems to be no time when death, if it is a misfortune, can be ascribed to its unfortunate subject. The third type of difficulty concerns the symmetry, mentioned above, between our attitudes to posthumous and prenatal existence. How can the former be bad if the latter is not?

It should be recognized that if these are valid objections to counting death as an evil, they will apply to many other supposed evils as well. The first type of objection is expressed in general form by the common remark that what you don't know can't hurt you. It means that even if a man is betrayed by his friends, ridiculed behind his back, and despised by people who treat him politely to his face, none of it can be counted as a misfortune for him so long as he does not suffer as a result. It means that a man is not injured if his wishes are ignored by the executor of his will, or if, after his death, the belief becomes current that all the literary works on which his fame rests were really written by his brother, who died in Mexico at the age of 28. It seems to me worth asking what assumptions about good and evil lead to these drastic restrictions.

All the questions have something to do with time. There certainly are goods and evils of a simple kind (including some pleasures and pains) which a person possesses at a given time simply in virtue of his condition at that time. But this is not true of all the things we regard as good or bad for a man. Often we need to know his history to tell whether something is a misfortune or not; this applies to ills like deterioration, deprivation, and damage. Sometimes his experiential s*tate* is relatively unimportant—as in the case of a man who wastes his life in the cheerful pursuit of a method of communicating with asparagus plants. Some-one who holds that all goods and evils must be temporarily assignable states of the person may of course try to bring difficult cases into line by pointing to the pleasure or pain that more complicated goods and evils cause. Loss, betrayal, deception, and ridicule are on this view bad because people suffer when they learn of them. But it should be asked how our ideas of human value would have to be constituted to accommodate these cases directly instead. One advantage of such an account might be that it would enable us to explain *why* the discovery of these misfortunes causes suffering—in a way that makes it reasonable. For the natural view is that the discovery of betrayal makes us unhappy because it is bad to be betrayed—not that betrayal is bad because its discovery makes us un-happy.

It therefore seems to me worth exploring the position that most good and ill fortune has as its subject a person identified by his history and his possibilities, rather than merely by his categorical state of the moment—and that while this subject can be exactly located in the sequence of places and times, the same is not necessarily true of the goods and ills that befall him.

2. Good

a. Release of the Soul from the Chains of the Body

Not all hold the view that death is bad. Some hold that death is good because it re-leases the soul from the chains of the body, brings happiness, and is to be met with content. We explore these positions now. We were introduced to ancient Greek philosopher Plato (ca. 427–348 BCE) in chapter 4, B.2.a. We consider his views again in this discussion of death as good. In the following selection from *Phaedo*, Plato recounts the death of his teacher and mentor, Socrates (ca. 470–399 BCE). As the story goes, Socrates was condemned to death by rulers in his society because he challenged views supported by the officials. While he had a chance to change his ways and go free, he chose to accept his conviction because he believed this was the right act to follow given his commitment to his society to follow the law (even when he disagreed with it). In what follows, Socrates provides insight into how he understands death. As he says, "I desire to prove to you that the real philosopher has reason to be of good cheer when he is about to die; and that after death he may hope to obtain the greatest good in the other world" (Plato 1953a, 414). On this view, death is good because it involves "the release of the soul from the chains of the body" (Plato 1953a, 418). This view of death is not simply a theoretical suggestion; it carries practical implications in that Socrates models for us how to approach death cheerfully. When faced with death, a death imposed by a government to which he remained loyal, Socrates says: "A prayer to the gods I may and must offer, that they will prosper my journey from this to the other world—even so—and so be it according to my prayer. Then he held his breath and drank off the poison quite readily and cheerfully" (Plato 195a, 476). The *Phaedo* gives us an early account of death as good.

Reading: *Phaedo* (Plato 1953a, 414, 417–19, 474, 475–76). Reprinted with the permission of Oxford Publishing Limited.

[Socrates] I desire to prove to you that the real philosopher has reason to be of good cheer when he is about to die; and that after death he may hope to obtain the greatest good in the other world. And how this may be, Simmias and Cebes, I will endeavour to explain. For I deem that the true votary of philosophy is likely to be misunderstood by other men; they do not perceive that of his own accord he is always engaged in the pursuit of dying and death; and if this be so, and he has had the desire of death all his life long, why when his time comes should he repine at that which he has been always pursuing and desiring?

Simmias said laughingly: Though I am not altogether in a laughing humour, you have made me laugh, Socrates; for I cannot help thinking that the many when they hear your words will say how truly you have described philosophers, and our people at home will likewise say that philosophers are in reality moribund, and that they have found them out to be deserving of the death which they desire.

And they are right, Simmias, in thinking so, with the exception of the words "They have found them out"; for they have not found either in what sense the true philosopher is moribund and deserves death, or what manner of death he deserves. But enough of them:—let us discuss the matter among ourselves. Do we attach a definite meaning to the word "death"?

To be sure, replied Simmias.

Is it not just the separation of soul and body? And to be dead is the completion of this; when the soul exists by herself and is released from the body, and the body is released from the soul. This, I presume, is what is meant by death?

Just so, he replied. . . .

". . . It has been proved to us by experience that if we would have pure knowledge of anything we must be quit of the body—the soul by herself must behold things by themselves; and then we shall attain that which we desire, and of which we say that we are lovers—wisdom; not while we live, but, as the argument shows, only after death; for if while in company with the body the soul cannot have pure knowledge, one of two things follows—either knowledge is not to be attained at all, or, if at all, after death. For then, and not till then, the soul will be departed from the body and exist by herself alone. In this present life, we think that we make the nearest approach to knowledge when we have the least possible intercourse or communion with the body, and do not suffer the contagion of the bodily nature, but keep ourselves pure until the hour when god himself is pleased to release us. And this getting rid of the foolishness of the body we may expect to be pure and hold converse with the pure, and to know of ourselves all that exists in perfection unalloyed, which, I take it, is no other than the truth. For the impure are not permitted to lay hold of the pure." These are the sort of words, Simmias, which the true lovers of knowledge cannot help saying to one another, and thinking. You would agree; would you not?

Undoubtedly, Socrates.

But, O my friend, if this be true, there is great reason to hope that, going whither I go, when I have come to the end of my journey I shall fully attain that which has been the pursuit of our lives. And therefore I accept with good hope this change of abode which is now enjoined upon me, and not I only, but every other man who believes that his mind has been made ready and that he is in a manner purified.

Certainly, replied Simmias.

And does it not follow that purification is nothing but that separation of the soul from the body, which has for some time been the subject of our argument; the habit of the soul gathering and collecting herself into herself from all sides out of the body; the dwelling in her place alone, as in another life, so also in this, as far as she can;—the release of the soul from the chains of the body? . . .

And this separation and release of the soul from the body is termed death? . . .

And the true philosophers, and they only, are ever seeking to release the soul. Is not the separation and release of the soul from the body their especial study?

That is true.

And, as I was saying at first, there would be a ridiculous contradiction in men studying to live as nearly as they can in a state like that of death, and yet repining when death comes upon them.

Clearly.

In fact, the true philosophers, Simmias, are always occupied in the practice of dying, wheretofore also to them least of all men is death terrible. Look at the matter thus:—if they have been in every way estranged from the body, and are wanting to be alone with the soul, when this desire of theirs is being granted, how inconsistent would they be if they trembled and repined, instead of rejoicing at their departure to that place, where, when they arrive, they hope to gain that which in life they desired—and their desire was for wisdom—and at the same time to be rid of the company of their enemy. Many a man who has lost by death an earthly love, or wife, or son, has been willing to go in quest of them to the world below, animated by the hope of seeing them there and of being with those for whom he yearned. And will he who is a true lover of wisdom, and is strongly persuaded in like manner that only in the world below he can worthily enjoy her, still repine at death? Will he not depart with joy? Surely he will, O my friend, if he be a true philosopher. For he will have a firm conviction that there, and there only, he can find wisdom in her purity. And if this be true, he would be very absurd, as I was saying, if he were afraid of death.

He would indeed, replied Simmias.

And when you see a man who is repining at the approach of death, is not his reluctance a sufficient proof that after all he is not a lover of wisdom, but a lover of the body, and is probably at the same time a lover of either money or power, or both?

Quite so, he replied. . . .

[*Ed. note: here, at the end of the* Phaedo, *Plato recounts Socrates's death.*]

. . . Soon I must drink the poison; and I think that I had better retire to the bath first, in order that the women may not have the trouble of washing my body after I am dead. . . .

. . . He arose and went into the chamber to bathe; Crito followed him and told us to wait. So we remained behind, talking and thinking of the subject of discourse, and also of the greatness of our loss; he was like a father of whom we were being bereaved, and we were about to pass the rest of our lives as orphans. When he had taken his bath his children were brought to him—(he had two young sons and an elder one); and the women of his family also came, and he talked to them and gave them a few directions in the presence of Crito; then he dismissed them and returned to us.

Now the hour of sunset was near, for a good deal of time had passed while he was within. When he came out, he sat down with us again after his bath, but not much was said. Soon the jailer, who was the servant of the Eleven, entered and stood by him saying:—To you, Socrates, whom after your time here I know to be the noblest and gentlest and best of all who ever came to this place, I will not impute the angry feelings of other men, who rage and swear at me, when, in obedience to the authorities, I bid them drink the poison—indeed, I am sure that you are not angry with me; for others, as you are aware, and not I, am to blame. And so fare you well, and try to bear lightly what must needs be—you know my errand. Then bursting into tears he turned and started on his way out.

Socrates looked up at him and said: I return your good wishes, and will do as you bid. Then turning to us, he said, How charming the man is: since I have been in prison he has always been coming to see me, and at times he would talk to me, and was as good to me as he could be, and see how generously he sorrows on my account. We must do as he says, Crito; and therefore let the cup be brought, if the poison is prepared; if not, let the attendant prepare some.

But said Crito, the sun is still upon the hill-tops, and is not yet set. I know that many a one takes the draught quite a long time after the announcement has been made to him, when he has eaten and drunk to his satisfaction and enjoyed the society of his chosen friends; do not hurry, there is enough time.

Socrates said: Yes, Crito, and therein they of whom you speak act logically, for they think that they will be gainers by the delay; but I likewise act logically in not following their example, for I do not think that I should gain anything by drinking the poison a little later; I should only be ridiculous in my own eyes for sparing and saving a life which is already down to its dregs. Please then to do as I say, and not refuse me.

Crito made a sign to the servant, who was standing by; and he went out, and having been absent for some time, returned with the jailer carrying the cup of poison. Socrates said: You, my good friend, who are experienced in these matters, shall give me directions how I am to proceed. The man answered: You have only to walk about until your legs are heavy, and then to lie down, and the poison will act. At the same time he handed the cup to Socrates, who in the easiest and gentlest manner, without the least fear or change of colour or features, and looking at the man sideways with that droll glance of his, took the cup and said: What do you say about making a libation out of the cup to any god? May I, or not? The man answered: We only prepare, Socrates, just so much as we deem enough. I understand, he said: but a prayer to the gods I may and must offer, that they will prosper my journey from this to the other world—even so—and so be it according to my prayer. Then he held his breath and drank off the poison quite readily and cheerfully.

b. More Happiness Than That of a King

Chuang Tzu (or Zhuangzi) (ca. 369–286 BCE) was, along with Lao-Tzu (ca. 575 BCE), one of the early, well-known Chinese philosophers who contributed to the development of philosophical Taoism (or Daoism), or school of the Way. It is believed that Chuang Tzu lived in Meng in Anhui in the state of Song, China. Chuang Tzu is known for his sharp wit on topics that typically are difficult to address, such as death. His teachings were penned by his followers in narrative form and handed down in history. For Chuang Tzu, death can be seen as good. In his parable portraying a passerby speaking with a skull found on the roadside, the passerby asks the skull whether it would want to come to life again. Note how the skull responds: "The skull frowned severely, wrinkling up its brow. 'Why would I throw away more happiness than that of a king on a throne and take on the troubles of a human being again?' it [the skull] said" (Chuang Tzu 1964, 115). In the passage that follows, Chuang Tzu speaks of the goodness or joy of death.

Reading: "Supreme Happiness," in *Basic Writings* (Chuang Tzu 1964, 114–15). Copyright 1964. Reprinted with the permission of Columbia University Press.

When Chuang Tzu went to Ch'u, he saw an old skull, all dry and parched. He poked it with his carriage whip and then asked, "Sir, were you greedy for life and forgetful of reason, and so came to this? Was your state overthrown and did you bow beneath the ax, and so came to this? Did you do some evil deed and were you ashamed to bring disgrace upon your parents and family, and so came to this? Was it through the pangs of cold and hunger that you came to this? Or did your springs and autumns pile up until they brought you to this?"

When he had finished speaking, he dragged the skull over and, using it for a pillow, lay down to sleep.

In the middle of the night, the skull came to him in a dream and said, "You chatter like a rhetorician and all your words betray the entanglements of a living man. The dead know nothing of these! Would you like to hear a lecture on the dead?"

"Indeed," said Chuang Tzu.

The skull said, "Among the dead there are no rulers above, no subjects below, and no chores of the four seasons. With nothing to do, our springs and autumns are as endless as heaven and earth. A king facing south on his throne could have no more happiness than this!"

Chuang Tzu couldn't believe this and said, "If I got the Arbiter of Fate to give you a body again, make you some bones and flesh, return you to your parents and family and your old home and friends, you would want that, wouldn't you?"

The skull frowned severely, wrinkling up its brow. "Why would I throw away more happiness than that of a king on a throne and take on the troubles of a human being again?" it said.

c. To Be Met with Content

Roman emperor Marcus Aurelius Antoninus (121–80 CE) is known in philosophy for his contributions to Stoic thought. His *Meditations* may be read as a series of practical philosophical exercises. Central to these exercises are instructions to analyze one's own judgments. During his reign, Aurelius faced numerous challenges in ensuring the welfare of his citizens. Constant warfare on the frontier of his empire and severe pestilence in society led him to confront death on a daily basis. His study of Stoic philosophy is evident in his thinking about death. According to Aurelius, we are to be "content" with death. As he says, "Do not despise death, but be well content with it, since this too is one of those things which nature wills. For such as it is to be young and to grow old, and to increase and to reach maturity, and to have teeth and beard and gray hairs, and to beget, and to be pregnant, and to bring forth, and all other natural operations which the seasons of thy life bring, such also is dissolution" (Aurelius 1909 [9.3]). What follows is a series of reflections on the goodness of death drawn from the meditations of the Roman emperor Aurelius.

Reading: *The Meditations of Marcus Aurelius* (Aurelius 1909 [4.5; 4.48; 6.28; and 9.3])

[4.5]
Death is such as generation is, a mystery of nature; a composition out of the same elements, and a decomposition into the same; and altogether not a thing of which any man should be ashamed, for it is not contrary to [the nature of] a reasonable animal, and not contrary to the reason of our constitution.

[4.48]

Think continually how many physicians are dead after often contracting their eyebrows over the sick; and how many astrologers after predicting with great pretensions the death of others; and how many philosophers after endless discourses on death or immortality. . . . Add to the reckoning all whom thou hast known, one after another. One man after burying another has been laid out dead, and another buries him; and all this in a short time. To conclude, always observe how ephemeral and worthless human things are, and what was yesterday a little mucus, tomorrow will be a mummy or ashes. Pass then through this little space of time comfortably to nature, and end thy journey in content, just as an olive falls off when it is ripe, blessing nature who produced it, and thanking the tree on which it grew.

[6.28]

Death is the cessation of the impressions through the senses, and of the pulling of the strings which move the appetites, and of the discursive movements of the thoughts, and of the service to the flesh.

[9.3]

Do not despise death, but be well content with it, since this too is one of those things which nature wills. For such as it is to be young and to grow old, and to increase and to reach maturity, and to have teeth and beard and gray hairs, and to beget, and to be pregnant, and to bring forth, and all other natural operations which the seasons of thy life bring, such also is dissolution. This, then, is consistent with the character of a reflecting man, to be neither careless nor impatient nor contemptuous with respect to death, but to wait for it as one of the operations of nature. As thou now waitest for the time when the child shall come out of thy wife's womb, so be ready for the time when thy soul shall fall out of this envelope.

C. REFLECTIONS: PAPERS, CONTACTS, AND LAST LETTER

1.(*) Public discourse typically supports the view that death is bad. Reflect on ways in which death is bad. What reasons are given for this view? Draw on your own personal experience and provide detail in the analysis.

2.(*) For many of us, it is difficult to think that death is good. Nevertheless, reflect on ways in which death is good. What reasons are given for this view? Draw on your own readings as well as your own personal experience and provide detail in the analysis.

3.   This is a continuation of your advance planning in preparation for death.

   a. In planning for death, it is not enough simply to sign a bunch of documents listing your wishes. You need to leave such documents where your heirs can find them. In this spirit, consider the following list of items that need to be available to your heirs should you die. If the document is relevant in your life, do you know where it is? Answer "yes," "no," or "NA" [not applicable], and indicate where they can be found. Indicate who knows where such documents can be found.

   i.   your paystubs (from the last year)
   ii.   utility bills (telephone, gas, electric) (from the last year)
   iii. bank statements (from the last year)
   iv.   investment statements (from the last year)
   v.   income tax returns (from the last three years)
   vi.   medical bills (from the last three years)
   vii. annual investment statements (from the last three years)
   viii. home improvement records (from the last three years)
   ix.   record of satisfied loans (from the last seven years)
   x.   contracts
   xi.   insurance documents
   xii. stock certificates and records
   xiii. property records
   xiv. records of pension or retirement plans
   xv.   marriage license
   xvi. birth certificate
   xvii. will
   xviii. adoption papers
   xix. death certificates of your loved ones
   xx.   records of paid mortgages
   xxi. list of key contacts (executor, durable medical power of attorney, lawyer, real estate agent, financial adviser, key family members, close friends)
   xxii. passwords to financial and personal websites

   b. A "last letter" became popular when soldiers began to leave "goodbye" letters for family members at the end of the twentieth century. The popularity of last letters also seems to correlate with the extension of the life span and individuals leaving words of wisdom and expressions of thanks to loved ones (Willis 2014). Your task is to write a "last letter" that is to be read by your loved one after your death. Choose a significant other and pen your parting thoughts.

D. FURTHER READINGS

Augustine. 2000. *The City of God.* Translated by Marcus Dods. New York: Modern Library.

Barry, Vincent. 2007. *Philosophical Thinking about Death and Dying.* Belmont, CA: Thomson Wadsworth.

Brennan, Samantha, and Robert J. Stainton, eds. 2009. *Philosophy and Death: Introductory Readings.* Toronto, Ontario: Broadview Press.

Chaudhuri, Saabira. 2011. "The 25 Documents You Need before You Die." *Wall Street Journal,* July 2. http://www.wsj.com/articles/SB1000142405270230362710457641023403925 8092.

Chidester, David. 2002. *Patterns of Transcendence: Religion, Death, and Dying.* 2nd ed. Belmont, CA: Wadsworth.

Harvey, Peter. 1990. *An Introduction to Buddhism.* Cambridge: Cambridge University Press.

Kushner, Harold. 1978. *When Bad Things Happen to Good People.* New York: Random House.

Luper, Steven. 2009. *The Philosophy of Death.* Cambridge: Cambridge University Press.

Rahula, Walpola. 1974. *What the Buddha Taught.* New York: Grove Press.

Willis, Claire. 2014. *Lasting Words: A Guide to Finding Meaning toward the Close of Life.* Brattelboro, VT: Green Writers Press.

— Chapter 10 —

To Be Feared or Not

A. CONTEXT

There is a significant body of reflection in the philosophical and psychological literature regarding whether death is to be feared. Perhaps the classic line by Woody Allen says it best: "I'm not afraid of death. I just don't want to be there when it happens." Those who fear death may base this reaction on a number of factors, including fear of the unknown, loss of control, fear of pain, loss of dignity, fear associated with the death of a loved one, or the implications brought about by sanctions supported by their religious traditions. Those who do not fear death may hold that, since one cannot experience death, one cannot know it and thus cannot and should not fear it. They may also hold that worrying about something that one has no control over doesn't do any good, so why worry over that which one cannot control? Still for others, there are reasons both to fear death and not to fear death based on life experiences. Whether one fears death turns on a host of considerations regarding how one views death, how one experiences the loss of another, and how one integrates cultural and spiritual traditions into one's personal views.

In the previous chapters, we have seen examples of those who fear death and those who do not. When death is understood as a gateway to "hell," "fire," or "attachment," it is typically feared. When it is understood as a gateway to "heaven," "paradise," or "liberation," it is typically not feared, unless, of course, one fears the unknowns associated with such states. When death is understood as "despair" and "loss," it is typically feared, and when it is understood as "hope" or "release from the chains of the body," it is typically not feared. The following selections offer additional accounts of whether death is to be feared or not. Selections from *The Tibetan Book of the Dead*, Canadian cultural anthropologist Ernest Becker, and US psychologists Tom Pyszczynski, Sheldon Solomon, and Jeff Greenberg illustrate reasons to fear death. Selections from the Taoist philosopher Lao-Tzu, the ancient Stoic philosopher Epicurus, a Peruvian author retelling the death of the fifteenth-century leader Pachacuti Inca Yupanqui, and a Native American Cherokee poet illustrate reasons not to fear death.

B. PERSPECTIVES

1. To Be Feared

a. Confronting One's Fears

As stated in chapter 4, B.1.c, *The Tibetan Book of the Dead* (*Bardo Thodol Chenmo*) is an eighth-century book reportedly written by the spiritual leader Padmasambhava. The book is a self-help guidebook intended to help the dying and recently deceased souls to find their way through the difficult stages of the afterlife. While the guide does not promote the view that we should fear death, it recognizes that we do fear death as it guides the dying person on how not to fear death. In the text, "bardos" stand for existential gaps in life's journey that must be addressed in order to reach liberation. Bardos are represented in terms of six realms of psychological states experienced by a dying person. To move through the bardos, an individual needs to face his or her fears that arise in life's journey. For instance, as the guidebook directs:

> Now when the bardo of dharmatā [i.e., "suchness"] dawns upon me,
> I will abandon all thoughts of fear and terror,
> I will recognize whatever appears as my projection
> and know it to be a vision of the bardo;
> now that I have reached this crucial point,
> I will not fear the peaceful and wrathful ones, my own projections.
> (*Tibetan Book of the Dead* 1992, 237)

While the goal is to overcome one's fears of various aspects of life, one may not accomplish this, and therefore one experiences rebirth as opposed to enlightenment. What follows are the main verses of the six bardos that guide a dying individual to face his or her fears of death.

Reading: "The Main Verses of the Six Bardos," in *The Tibetan Book of the Dead* (1992, 235–38)

Now when the bardo of birth is dawning upon me,
I will abandon laziness for which life has no time,
enter the undistracted path of study, reflection and meditation,
making projections and the mind the path, and realize the three key *kayas* [i.e.,
   bodies of Buddha's enlightenment, corresponding to mind, speech, and body];
now that I have once attained a human body,
there is no time on the path for the mind to wander.

Now when the bardo of dreams is dawning upon me,
I will abandon the corpse-like sleep of careless ignorance,
and let my thoughts enter their natural state without distraction;
controlling and transforming dreams in luminosity,
I will not sleep like any animal
but unify complete sleep and practice.

Now when the bardo of samādhi-meditation [an advanced level of meditation]
   dawns upon me,
I will abandon the crowd of distractions and confusions,
and rest in the boundless state without grasping or disturbance;
firm in the two practices: visualization and complete,
at this time of meditation, one-pointed, free from activity,
I will not fall into the power of confused emotions.

Now when the bardo of the moment before death dawns upon me,
I will abandon all grasping, yearning and attachment,
enter undistracted into clear awareness of the teaching,
and eject my consciousness into the space of unborn mind;
as I leave this compound body of flesh and blood
I will know it to be a transitory illusion.

Now when the bardo of dharmatā [i.e., "suchness"] dawns upon me,
I will abandon all thoughts of fear and terror,
I will recognize whatever appears as my projection
and know it to be a vision of the bardo;
now that I have reached this crucial point,
I will not fear the peaceful and wrathful ones, my own projections.

Now when the bardo of becoming dawns upon me,
I will concentrate my mind one-pointedly,
and strive to prolong the results of good karma,
close the womb-entrance and think of resistance;
this is the time when perseverance and pure thought are needed,
abandon jealousy, and meditate on the guru with his consort.

With mind far off, not thinking of death's coming,
performing these meaningless activities,
returning empty-handed now would be complete confusion;
the need is recognition, holy dharma [i.e., eternal truth],
so why not practice dharma at this very moment?
From the mouths of siddhas [i.e., ascetics who have attained enlightenment]
    come these words:
If you do not keep your guru's teaching in your heart
will you not become your own deceiver?

b. A Potential for Terror

Cultural anthropologist Ernest Becker (1924–74) is known for his account of the ways in which humans deny death. Well before the formation of the discipline of gerontology (i.e., the study of aging) and more open discussions of death, Becker spoke directly about what he called the "terror of death," which results from the basic biological need to control the self and its impending death. In 1974, he was awarded a Pulitzer Prize for his work in *The Denial of Death*. In *The Denial of Death*, Becker seeks to provide "a network of arguments based on the universality of the fear of death, or 'terror' as I prefer to call it, in order to convey how all-consuming it is when we look it full in the face" (1973, 15). According to Becker, the terror of death is pervasive and so profound in human life that humans conspire to keep it unconscious. In keeping it unconscious and striving to hide the inevitability of death, humans live a vital lie and fail to come to terms with their inevitable nature, thus creating in their lives multiple and varied problems in living. A selection from *The Denial of Death* follows.

Reading: "The 'Healthy-Minded' and 'Morbidly-Minded' Arguments," in *The Denial of Death* (Becker 1973, 11–12, 13, 14, 15). Reprinted with the permission of Free Press, a division of Simon and Schuster, Inc. Copyright 1974 by The Free

The first thing we have to do with heroism is to lay bare its underside, show what gives human heroics its specific nature and impetus. Here we introduce directly one of the great rediscoveries of modern thought: that of all things that move man, one of the principal ones is his terror of death. After Darwin the problem of death as an evolutionary one came to the fore, and many thinkers immediately saw that it was a major psychological problem for man. They also very quickly saw what real heroism was about, as Shaler writes just at the turn of the century: heroism is first and foremost a reflex of the terror of death. We admire most the courage to face death; we give such valor our highest and most constant adoration; it moves us deeply in our hearts because we have doubts about how brave we ourselves would be. When we see a man bravely facing his own extinction we rehearse the greatest victory we can imagine. And so the hero has been the center of human honor and acclaim since probably the beginning of specifically human evolution. But even before that our primate ancestors deferred to others who were extra powerful and courageous and ignored those who were cowardly. Man has elevated animal courage into a cult. . . .

The "Healthy-Minded" Argument

There are "healthy-minded" persons who maintain that fear of death is not a natural thing for man, that we are not born with it. An increasing number of careful studies on how actual fear of death develops in the child agree fairly well that the child has no knowledge of death until about the age of three to five. How could he? It is too abstract an idea, too removed from his experience. He lives in a world that is full of living, acting things, responding to him, amusing him, feeding him. He doesn't know what it means for life to disappear forever, nor theorize where it would go. Only gradually does he recognize that there is a thing called death that takes some people away forever; very reluctantly he comes to admit that it sooner or later takes everyone away, but this gradual realization of the inevitability of death can take up until the ninth or tenth year.

As we see later on, this view is very popular today in the widespread movement toward unrepressed living, the urge to a new freedom for natural biological urges, a new attitude of pride and joy in the body, the abandonment of shame, guilt, and self-hatred. From this point of view, fear of death is something that society creates and at the same time uses against the person to keep him in submission; the psychiatrist Moloney talked about it as a "culture mechanism," and Marcuse as an "ideology." Norman O. Brown, in a vastly influential book that we shall discuss at some length, went so far as to say that there could be a birth and development of the child in a "second innocence" that would be free of the fear of death because it would not deny natural vitality and would leave the child fully open to physical living. . . .

The "Morbidly-Minded" Argument

The "healthy-minded" argument just discussed is one side of the picture of the accumulated research and opinion on the problem of the fear of death, but there is another side. A large body of people would agree with these observations on early experience and would admit that experiences may heighten natural anxieties and later fears, but these people would also claim very strongly that nevertheless the fear of death is natural and is present in everyone, that it is the basic fear that influences all others, a fear from which no one is immune, no matter how disguised it may be. William James spoke very early for this school, and with his usual colorful realism he called death "the worm at the core" of man's pretensions to happiness. No less a student of human nature than Max Scheler thought that all men must have some kind of certain intuition of this "worm at the core," whether they admitted it or not. Countless other authorities . . . belong to this school: students of the stature of Freud, many of his close circle, and serious researchers who are not psychoanalysts. . . .

I frankly side with this second school—in fact, this whole book is a network of arguments based on the universality of the fear of death, or "terror" as I prefer to call it, in order to convey how all-consuming it is when we look it full in the face.

c. A Constant Threat

US social psychologists Tom Pyszczynski, Sheldon Solomon, and Jeff Greenberg are the originators of "terror management theory," an existential psychological model that helps explain why humans react the way they do to the threat of death, and how this reaction influences their post-threat cognition and emotion. In *In the Wake of 9/11: The Psychology of Terror,* Pyszczynski and his colleagues analyze the roots of terrorism and American reactions to the attacks on the World Trade Center and the Pentagon on September 11, 2001. The authors focus primarily on the reactions to the 9/11 attacks, but their model is applicable to all instances of terrorism. Simply put, humans develop ways to manage terror, horror, and dread in order to mitigate their effects on how they live their lives. Terror management theory tells us something about how we respond to death and the reality that we all die. In what follows, Pyszczynski and his colleagues address how death is a constant threat in human life. Such threat is mitigated, they say, because "cultures provide ways to view the world—worldviews—that 'solve' the existential crisis engendered by the awareness of death. Cultural worldviews consist of humanly constructed beliefs about the nature of reality that are shared by individuals in a group that function to mitigate the horror and blunt the dread caused by knowledge of the reality of the human condition, that we all die" (2003, 16). Note in what follows how Pyszczynski and his colleagues draw from Kierkegaard (see chapter 8, B.1) in developing their analysis.

Reading: "Grave Matters: On the Role of Death in Life," in *In the Wake of 9/11: The Psychology of Terror* (Pyszczynski, Solomon, and Greenberg 2003, 13–16). Copyright © 2003 American Psychological Association. Reproduced with permission.

Accordingly, our analysis of the human condition commences with the basic Darwinian assumption that all living things share a biological predisposition toward self-preservation, because such a tendency facilitates staying alive long enough to reproduce and pass one's genes on to future generations. We view this as a very basic adaptation that arose early in the history of life, long before humans, primates, or even mammals emerged as species. Different forms of life vary immensely, however, in terms of the structural, functional, behavioral, and (in some cases) psychological adaptations they have acquired in the context of millions (and in some cases billions) of years of evolution to render them fit for the specific environmental niches that they inhabit. Plants derive their sustenance by converting solar energy into carbon dioxide through photosynthesis; spiders construct elaborate silk webs in which they catch their food; turtles parry the thrusts of their predatory opposition by retreating into the bony or leathery shells erected on their backs; wolves travel in packs in coordinated search for prey and regurgitate half-digested food to feed their young. These are just a few examples of the seemingly infinite variety of effective adaptations to the demands of physical reality that various life forms have acquired to facilitate the arduous task of staying alive and perpetuating their genetic material over time.

What about human beings? . . .

One especially important aspect of human mentality, recognized by early Greek and Roman philosophers and subsequently central to Kierkegaard's penetrating understanding of the human condition, is that human beings are conscious and self-conscious. Other creatures are sensate and perceptive, but only human beings are aware of their awareness (consciousness) and of themselves as potential objects of their subjective inquiry (self-conscious [*sic*] may be expressed in this way: "I am and I know that I am; I know that I know that I am; I know that I know that I know that I am"). Consciousness and self-consciousness have important cognitive and affective consequences. Cognitively, we become time-binding animals, able to reflect on the past and anticipate the future and, in so doing, enhance our prospects for survival in the present. Affectively, consciousness and self-awareness infuse us with awe and dread. To be alive and to know it is awesome: grand, breathtaking, tremendous, remarkable, amazing, astounding, and humbling. Every one of us, in our finer moments, is overwhelmed with the sheer joy of being alive. In Otto Rank's lovely expression, we are "the temporal representative of the cosmic primal force": ultimately descended from, and consequently directly related to, the first living organism, as well as everything that has ever been alive, is currently living, and will ever live in the future.

However, consciousness and self-awareness also necessarily engender dread: fear, trepidation, anxiety, alarm, fright, horror, and, in due course, unmitigated terror. First, knowing that one is alive and being able to anticipate the future inevitably produces the unsettling awareness of one's inexorable death. . . .

Second, people also recognize that death not only is unavoidable but also can often occur quite tragically and prematurely (relative to an individual's normal life span) for reasons that can never be adequately anticipated or controlled. Californians know that the next earthquake might shake them off the face of the earth; Indians know that the next tidal wave might wash them into the sea; skiers know that the next downhill run might end by turning their heads into Jackson Pollock paintings on the side of trees; we all know that our next stomachache may be a softball-sized malignant tumor, a harbinger of our impending demise. We have for a long time known but are now especially aware that our next plane trip, whether for business or pleasure, may not end happily; we also now know that the white dust that fell from the unexpected greeting card we just received in the mail may kill us; we know that life is precarious and that each of us is ultimately vulnerable and helpless in the wake of the virtually infinite number of mundane and arcane ways our existence might be terminated; we know.

Finally, we also know and are horrified by the realization that we are corporeal creatures—sentient pieces of bleeding, defecating, urinating, vomiting, exfoliating, perspiring, fornicating, menstruating, ejaculating, flatulence-producing, expectorating meat—that may ultimately be no more enduring or significant than cockroaches or cucumbers. The continuous awareness of these circumstances within which we live, faced with inevitable death, compounded by the recognition of tragedy magnified by our carnal knowledge makes us humans vulnerable to potentially overwhelming terror at virtually any given moment. Yet people rarely experience this existential terror directly.

What saves us is culture. Cultures provide ways to view the world—worldviews—that "solve" the existential crisis engendered by the awareness of death. Cultural worldviews consist of humanly constructed beliefs about the nature of reality that are shared by individuals in a group that function to mitigate the horror and blunt the dread caused by knowledge of the reality of the human condition, that we all die.

2. Not to Be Feared

a. Not Frightening

Not all hold the view that death is to be feared. Some hold the view that death is not to be feared. We explore this option now. Lao-Tzu (6th c. BCE), or Laozi, is a well-known sixth-century BCE Taoist, or Daoist, philosopher from the southern Chinese state of Ch'u. He is said to have been a contemporary of the Chinese philosopher Confucius (551–497 BCE), who taught his followers about the good life through a series of aphorisms or sayings. Lao-Tzu's work, the *Tao Te Ching* (or *Daodejing* or *Dao De Jing*), addresses the Tao or Dao, the nontranscendent source, ideal, or "way" of existence, which we cannot know truly because of our limited words and descriptions. His message encourages living in harmony and balance with the Tao, and accepting the ebbs and flows of life as well as death. As Lao-Tzu says about death, "Men come forth and live; they enter (again) and die" (1891, 50). Since death is part of life, it should not be feared. As he says, "The people do not fear death; to what purpose is it to (try to) frighten them with death?" (1891, 74). What follows is a selection from the *Tao Te Ching*, which encourages us to accept death as a part of life.

Reading: *Tao Te Ching* (Lao-Tzu 1891 [secs. 6, 33, 50, 74, 76])

6

The valley spirit [i.e., that which supports and gives forth life] dies not, aye the same;
The female mystery thus do we name.
Its gate, from which at first they issued forth,
Is called the root from which grew heaven and earth.

Long and unbroken does its power remain,
Used gently, and without the touch of pain.

33

He who knows other men is discerning; he who knows himself is intelligent. He who overcomes others is strong; he who overcomes himself is mighty. He who is satisfied with his lot is rich; he who goes on acting with energy has a (firm) will.

He who does not fail in the requirements of his position, continues long; he who dies and yet does not perish, has longevity.

50

Men come forth and live; they enter (again) and die.

Of every ten three are ministers of life (to themselves); and three are ministers of death.

There are also three in every ten whose aim is to live, but whose movements tend to the land (or place) of death. And for what reason? Because of their excessive endeavours to perpetuate life.

But I have heard that he who is skillful in managing the life entrusted to him for a time travels on the land without having to shun rhinoceros or tiger, and enters a host without having to avoid buff coat or sharp weapon. The rhinoceros finds no place in him into which to thrust its horn, nor the tiger a place in which to fix its claws, nor the weapon a place to admit its point. And for what reason? Because there is in him no place of death.

74

The people do not fear death; to what purpose is it to (try to) frighten them with death? If the people were always in awe of death, and I could always seize those who do wrong, and put them to death, who would dare to do wrong?

There is always One who presides over the infliction of death. He who would inflict death in the room of him who so presides over it may be described as hewing wood instead of a great carpenter. Seldom is it that he who undertakes the hewing, instead of the great carpenter, does not cut his own hands!

76

Man at his birth is supple and weak; at his death, firm and strong. (So it is with) all things. Trees and plants, in their early growth, are soft and brittle; at their death, dry and withered.

Thus it is that firmness and strength are the concomitants of death; softness and weakness, the concomitants of life.

Hence he who (relies on) the strength of his forces does not conquer; and a tree which is strong will fill the out-stretched arms, (and thereby invites the feller.)

Therefore the place of what is firm and strong is below, and that of what is soft and weak is above.

b. Nothing to Us

The ancient Stoic philosopher Epicurus (341–271 BCE) is one of the major philosophers of the Hellenistic period, the three centuries following the death of ancient Greek military commander Alexander the Great in 323 BCE and Greek philosopher Aristotle in 322 BCE. Epicurus teaches that the good life is characterized by *ataraxia,* or peace and freedom from fear, and *apnonia,* or the absence of pain. He teaches that, in order to live the good life, we should try to limit our desires, practice prudence in all matters, and eliminate the fear of the gods and of death ("Epicurus" 2017). For Epicurus, it is possible to eliminate the fear of death. Death should mean nothing to us because we do not experience it. As Epicurus says, "You should accustom yourself to believing that death means nothing to us, since every good and every evil lies in sensation; but death is the privation of sensation" (1963, 179). Put another way, where one is, death is not, and where death is, one is not. Death is neither good nor bad; it is neutral because when one is dead one lacks experience. As a consequence, there is no reason to fear death. What follows is a selection from a letter from Epicurus to his contemporary Menoeceus.

Reading: "Letter to Menoeceus" (Epicurus 1963, 179–81)

You should accustom yourself to believing that death means nothing to us, since every good and every evil lies in sensation; but death is the privation of sensation. Here a correct comprehension of the fact that death means nothing to us makes the mortal aspect of life pleasurable, not by conferring on us a boundless period of time but by removing the yearning for deathlessness. There is nothing fearful in living for the person who has really laid hold of the fact that there is nothing fearful in not living. So it is silly for a person to say that he dreads death—not because it will be painful when it arrives but because it pains him now as a future certainty; for that which makes no trouble for us when it arrives is a meaningless pain when we await it. This, the most horrifying of evils, means nothing to us, then, because so long as we are existent death is not present and whenever it is present we are nonexistent. Thus it is of no concern either to the living or to those who have completed their lives. For the former it is nonexistent, and the latter are themselves nonexistent.

Most people, however, recoil from death as though it were the greatest of evils; at other times they welcome it as the end-all of life's ills. The sophisticated person, on the other hand, neither begs off from living nor dreads not living. Life is not a stumbling block to him, nor does he regard not being alive as any sort of evil. As in the case of food he prefers the most savory dish to merely the larger portion, so in the case of time he garners to himself the most agreeable moments rather than the longest span.

Anyone who urges the youth to lead a good life but counsels the older man to end his life in good style is silly, not merely because of the welcome character of life but because of the fact that living well and dying well are one and the same discipline. Much worse off, however, is the person who says it were well not to have been born "but once born to pass through Hades' portals as swiftly as may be." Now if he says such a thing from inner persuasion why does he not withdraw from life? Everything is in readiness for him once he has firmly resolved on this course. But if he speaks facetiously he is a trifler standing in the midst of men who do not welcome him.

It should be borne in mind, then, that the time to come is neither ours nor altogether not ours. In this way we shall neither expect the future outright as something destined to be nor despair of it as something absolutely not destined to be.

c. Rest with My Father the Sun

Pachacuti Inca Yupanqui (or Pachacutec or Pachakutiq Inka Yupanki) was the ninth Inca ruler (1438–71/72) of the Kingdom of Cusco in Peru, which he transformed into the Inca Empire. Most archaeologists believe that the famous Inca site of Machu

Picchu, one of the major wonders of the world, was built as an estate for Pachacuti. During his reign, Cusco grew from a small area into a great empire. The Inca Empire located in Cusco was the largest empire in pre-Columbian America. The Inca civilization arose in Peru sometime in the early thirteenth century, and the last Inca stronghold was conquered by the Spaniards in 1572 ("Pachacuti" 2017).

The Inca people believed in Viracocha (also Pachacamac), who created all living things, and Pachamama, the wife of Viracocha and goddess of earth who protected agriculture. Death was a passage to the next world that was full of difficulties. The spirit of the dead, *camaquen*, would need to follow a long dark road to arrive at rest. Most Incas imagined the afterworld to be similar to the Euro-American notion of heaven, illuminated by the sun and flower-covered fields. It was important for the Inca people to ensure that they did not die as a result of burning and that the dead body was not cremated. This is because of the belief that a vital force would disappear and threaten their passage to the afterworld. Those who obeyed the Inca moral code—*ama suwa* (do not steal), *ama llulla* (do not lie), and *ama quella* (do not be lazy) —went to live in the Sun's warmth (or, as is said below, went to "rest with my father the Sun" [Sarmiento de Gamboa 1907, 138]). Those who did not follow the Inca moral code spent their eternal days on the cold earth. According to Inca belief, Inti is the Sun god and patron deity of the holy city of Cusco. What follows is a retelling of the death of the Inca ruler Pachacuti Inca Yupanqui.

Reading: "Death of Pachacuti Inca Yupanqui," in *History of the Incas* (Sarmiento de Gamboa 1907, 138–39)

Being in the highest prosperity and sovereignty of his life, he fell ill of a grave infirmity, and, feeling that he was at the point of death, he sent for all his sons who were then in the city. In their presence he first divided all his jewels and contents of his wardrobe. Next he made them plough furrows in token that they were vassals of their brother, and that they had to eat by the sweat of their hands. He also gave them arms in token that they were to fight for their brother. He then dismissed them.

He next sent for the Incas *orejones* of Cuzco, his relations, and for Tupac Inca his son to whom he spoke, with a few words, in this manner: "Son! you now see how many great nations I leave to you, and you know what labour they have cost me. Mind that you are the man to keep and augment them. No one must raise his two eyes against you and live, even if he be your own brother. I leave you these our relations that they may be your councillors. Care for them and they shall serve you. When I am dead, take care of my body, and put it in my houses at Patallacta. Have my golden image in the House of the Sun, and make my subjects, in all the provinces, offer up solemn sacrifice, after which keep the feast of *purucaya*, that I may go to rest with my father the Sun." Having finished his speech they say that he began to sing in a low and sad voice with words of his own language. They are as follows:

> "I was born as a flower of the field,
> As a flower I was cherished in my youth,
> I came to my full age, I grew old,
> Now I am withered and die."

Having uttered these words, he laid his head upon a pillow and expired, giving his soul to the devil, having lived a hundred and twenty-five years. For he succeeded, or rather he took the Incaship into his hands when he was twenty-two, and he was sovereign one hundred and three years.

d. Walking On

The Cherokee Indians are Native American people who migrated from the Great Lakes area of the United States and settled primarily in the Southeast and in the states of Georgia, North Carolina, South Carolina, and Tennessee. They are one of the first Native American people to become US citizens and constitute the largest federally recognized Native American tribe in the United States at the beginning of the twenty-first century. Today, they are the largest Indian nation in the United States. The Cherokees are organized in terms of a strong tribal identity, and their tribe is divided into smaller sections and led by chiefs. The Cherokee view of death expresses a belief in pantheism, the view that the universe or nature is identical with the divine, as well as an Abrahamic view of the divine (see chapter 5, B.2) inherited from European settlers. It also expresses the belief that the Great Spirit is feminine, an idea that differs from the Abrahamic view that the divine is masculine. As the poem "Walk On" says,

> And as the door of the Great Spirit world came closer
> my fear loomed up inside sometimes. . . .
>
> But something called me forth—
> the Morning Star rose with each day—
> and my prayer became a centering—and still I walked on,

until I began to hear the Song of the Mother,
and Her arms embraced me so,
that instead of walking She carried me right to the door.
And as the door opened, I heard her song,
and Her Song lifted me up, so I could soar.

("Walk On" 2003, 33)

The Cherokee Indian poem that follows teaches us not to fear death because what
awaits those who die will lift them up.

Reading: "Walk On—A Contemporary Cherokee Poem" ("Walk On" 2003, 32–33)

Good morning, Grandfather.
I entered this life a ways back
and put skin on to walk two-legged on this Creation—
and what a glorious time it was.

It taught me about breath
and about sense and feeling and caring through my heart.
And I walked on around that Red Road,
looking and trying to understand more
about the mystery and the secrets She holds.

And You spoke to me through the wind,
and You sang to me through the birds.
And you brought challenges forth so that
I might listen to the message You bring me more sincerely.
And I kept walking down this road.

And I came 'round the bend
at the middle of that curve in the road
and I began to find a secret in the Spirit of my Self. . . .

And still I walked on, sometimes blind and deaf,
and sometimes with pain.
But I fought with my fears and I embraced my unknowingness—
and still I walked on.
And my children and my family stood beside me

and we came to know each other in those later years more than we
had before—for some of our falseness had fallen away—
and still I walked on.

And I kept walking on this road towards You,
towards that other world that grew closer to me with each
     Step.
And as the door of the Great Spirit world came closer
my fear loomed up inside sometimes. . . .

But something called me forth—
the Morning Star rose with each day—
and my prayer became a centering—and still I walked on,
until I began to hear the Song of the Mother,
and Her arms embraced me so,
that instead of walking She carried me right to the door.
And as the door opened, I heard her song,
and Her Song lifted me up, so I could soar.

C. REFLECTIONS: FEAR ASSESSMENT

1.(*) Reflect on ways in which death is to be feared. What reasons are given for this view? Draw on your own personal experience and supplement the analysis with philosophical, clinical, and/or sacred text interpretations.

2.(*) Reflect on ways in which death is not to be feared. What reasons are given for this view? Draw on your own personal experience and supplement the analysis with philosophical, clinical, and/or sacred text interpretations.

3.   Reflect on whether you fear dying and death.
    a. Complete the assessment on the next page of your fear of dying and death. Check the column that applies.

Fear of Dying and Death Assessment

Fears	No	Sometimes	Yes
1. Do you fear dying?			
2. Do you fear that your dying will be painful?			
3. Do you fear a loss of physical or mental control when you are dying?			
4. Do you fear the medical interventions that might be used when you are dying?			
5. Do you fear that your dying will be expensive?			
6. Do you fear that your dying will be a burden on your loved ones?			
7. Do you fear that you will be lonely when you are dying?			
8. Do you fear that your dying will be long and drawn out?			
9. Do you fear death?			
10. Do you fear what happens after death?			
11. Do you fear leaving your loved ones when you die?			
12. Do you fear leaving your favorite possessions when you die?			
13. Do you fear having unresolved regrets when you die?			
14. Do you fear having uncompleted tasks when you die?			
15. Do you fear leaving the world you know when you die?			
16. Do you fear what you do not know about death?			

TOTALS

b. Given your responses (in 3.a), what can you say about whether you fear death? If you fear death, why and how do you fear death? If not, why not? How does your experience line up with views shared in this chapter? Be sure to give details in your reflections.

D. FURTHER READINGS

Becker, Ernest. 1997. *The Fear of Death*. New York: First Press Perspectives.

Byrock, Ira. 1998. *Dying Well*. New York: Riverhead Books.

Conte, H. R., M. B. Weiner, and R. Plutchik. 1982. "Measuring Death Anxiety: Conceptual, Psychometric, and Factor-Analytic Aspects." *Journal of Personality and Social Psychology* 43 (4): 775–85.

Feifel, Herman. 1959. *The Meaning of Death*. New York: McGraw-Hill.

Gehlek, Rimpoche Nawang 2001. *Good Life, Good Death*. New York: Riverhead Books.

Hanh, Thich Nhat. 2003. *No Death, No Fear: Comforting Wisdom for Life*. New York: Riverhead.

Yalom, Irvin D. 2009. *Staring at the Sun: Overcoming the Terror of Death*. New York: Jossey-Bass.

— Chapter 11 —

To Be Grieved and How

A. CONTEXT

There is a significant body of literature on death and dying regarding how death is to be grieved. In fact, one might say that one of the most important topics in discussions on death is grief or bereavement. Often individuals who have lost a significant other and experience grief seek out literature or a discussion group on grief and its resolution. While some recommendations regarding how to grieve are shared, recommendations are diverse, ranging from acceptance to rage, the very cognitive to the very emotional, and the very private to the very public.

One might keep in mind that grief is a human response not simply to death but to all kinds of losses. Whether one loses a loved one, a pet, a favorite possession, or a cherished reputation, humans respond to such losses, and there are cognitive and emotional patterns in such responses. Such responses can differ widely in degrees and kinds. Anyone who has experienced the loss of a loved one knows that this type of loss can be far more intense than the loss of a favorite article of clothing. Nevertheless, in some sense, loss is loss.

What follows is a range of responses regarding how death is to be grieved from a variety of traditions of thought. An Australian folktale captures how death is to be met with rage. Homer's story of Odysseus visiting the Kingdom of the Dead to see his deceased mother illustrates how death is to be met with a pierced heart. A selection from the *Sutta Nipata* of the *Sutta Pitaka* addresses how death is to be met with acceptance. The "Parable of the Mustard Seed" from foundational texts of

Theravada Buddhism illustrates how death can be approached calmly after one recognizes its universal occurrence. The Greek philosopher Epictetus discusses how death is not to be grieved. The British philosopher and theologian C. S. Lewis shares his struggle to find resolution of his grief after the loss of his beloved wife. Swiss-born psychiatrist Elisabeth Kübler-Ross provides a contemporary Western clinical account of common responses to death through what has come to be known as "the five stages of grief." Vanderbilt University nurse Kelley Allen explores culturally diverse responses in the bereavement process in the context of grieving the loss of a child in the neonatal intensive care unit. Taken together, the selections that follow provide a sampling of views regarding how humans grieve in the face of death.

B. PERSPECTIVES

1. To Be Met with Rage

In the oral traditions of Australia, the myth of the Dreamtime plays an important role. According to Aboriginal beliefs that date back ten thousand years, "Dreamtime" was the ancient time in which the first ancestors were created and acquired their basic knowledge and power (Dixon 1996, 130). In the Dreamtime myth, ancestors shaped the world as a place for humans to exist. After death, they left eternal traces in the landscape that they had helped construct (Chidester 2002, 43). In the myth of Dreamtime, death is thought to originate because of mistakes made by humans. One such story is the tale of Jinini. In the story, Jinini, son of mother Bima and father Purukapali, who were also wife and husband, dies because of a careless act of Bima and her lover, Japara. The story shares how the scorned Purukapali decrees that "as his son had died, so must the whole creation die" (28). It also illustrates how Purukapali reacts to the death of Jinini. As the story goes, "When the father heard of the tragedy, he was demented with rage and grief" (28). The story that follows explains the origin of death in the Australian Dreamtime myth and how the death of a son leads to inconsolable grief.

Reading: "The Death of Jinini—An Australian Aboriginal Tale," in *The Dreamtime: Australian Aboriginal Myths in Paintings by Ainslie Roberts* (Mountford 1965, 28)

When the world was young there was no death. This calamity was brought about by the wrongdoings of the woman Bima, and her lover Japara.

Bima had a son, Jinini, of whom her husband, Purukupali, was very fond. Every morning when Bima set out to collect food she took Jinini with her, and every evening she brought him back to his doting father, together with the food she had gathered.

An unmarried man, Japara, who lived in the same camp, constantly followed Bima and persuaded her to leave Jinini under the shade of a tree while she accompanied him into the jungle. This intrigue had been going on for some time when, on one very hot day, Bima and her lover stayed away too long. When the mother returned, she saw, to her horror, that the shade of the tree had moved and Jinini was lying dead in the blazing sun.

When the father heard of the tragedy, he was demented with rage and grief. He punished his wife severely for her carelessness; then, picking up the dead body of Jinini, Purukupali walked into the sea and drowned himself. At the same time he decreed that, as his son had died, so must the whole creation die, never to come to life again. And so it has remained from those remote times until now.

2. To Be Met with a Pierced Heart

The *Odyssey* by Greek poet Homer is a well-known tale. It was composed near the end of the eighth century BC, somewhere in Ionia, the Greek coastal region of Anatolia. The poem centers on the Greek hero Odysseus and his journey home after the fall of Troy, an ancient city located in what is now known as Anatolia in modern Turkey. It takes Odysseus ten years to reach Ithaca after the ten-year Trojan War. In his absence, it is assumed that he has died, and his wife Penelope and son Telemachus must deal with a group of unruly suitors who compete for Penelope's hand in marriage. After many years and many challenges, Odysseus arrives home to his wife and son and returns his kingdom to order ("Homer" 2017).

In book 11 of the *Odyssey*, Odysseus journeys to the Kingdom of the Dead or Hades, where his mother, Anticleia, along with others, is found. Odysseus speaks with the Theban prophet Tiresias, who foretells Odysseus's fate—that he will return home, will reclaim his wife and palace from the suitors, and then will make another trip to a distant land to appease Poseidon, angry brother of Zeus and god of the sea. When Tiresias departs for the Kingdom of the Dead, Odysseus calls other spirits toward him. He speaks with his mother, Anticleia, who updates him on the affairs of Ithaca and relates how she died of grief waiting for his return. As Anticleia explains, "'And this is how I sickened and died. The Archeress did not shoot me in my own house with those gentle shafts that never miss; it was no disease that made me pine away: but I missed you so much, and your clever wit and your gay merry ways, and life was sweet no longer, so I died'" (Homer 1937, 128). In what follows, Homer provides a classic retelling of how death is met with a pierced and grieving heart.

Reading: "How Odysseus Visited the Kingdom of the Dead," in *The Odyssey* (Homer 1937, 125, 126, 127, 128, 137 [bk. 11])

"Then came the soul of Theban Teiresias [i.e., guardian of Hadês, or hell], holding a golden rod. He knew me, and said, 'What brings you here, unhappy man, away from the light of the sun, to visit this unpleasing place of the dead? Move back from the pit, hold off your sharp sword, that I may drink of the blood and tell you the truth.'

"As he spoke, I stept back from the pit, and pushed my sword into the scabbard. He drank of the blood, and only then spoke as the prophet without reproach."

[Odysseus] ". . . [']But here is something I want to ask, if you will explain it to me. I see over there the soul of my dead mother. She remains in silence near the blood, and she would not look at the face of her own son, or say a word to him. Tell me, prince, how may she know me for what I am?'

"He answered, 'I will give you an easy rule to remember. If you let any one of the dead come near the blood, he will tell you what is true; if you refuse, he will go away.'

"When he had said this, the soul of Prince Teiresias returned into the house of Hadês, having uttered his oracles. But I stayed where I was until my mother came near and drank the red blood. At once she knew me, and made her meaning clear with lamentable words: . . .

[Odysseus wonders how his mother died.]

[Mother:] "'And this is how I sickened and died. The Archeress did not shoot me in my own house with those gentle shafts that never miss; it was no disease that made me pine away: but I missed you so much, and your clever wit and your gay merry ways, and life was sweet no longer, so I died.'

[Odysseus:] "When I heard this, I longed to throw my arms round her neck. Three times I tried to embrace the ghost, three times it slipt through my hands like a shadow or a dream. A sharp pang pierced my heart, and I cried out straight from my heart to hers:

"'Mother dear! Why don't you stay with me when I long to embrace you? Let us relieve our hearts, and have a good cry in each other's arms. Are you only a phantom which awful Persephoneia [i.e., daughter of Zeus] has sent to make me more unhappy than ever?'

"My dear mother answered:

"'Alas, alas, my child, most luckless creature on the face of the earth! Persephoneia is not deceiving you, she is the daughter of Zeus [i.e., ancient Greek god and supreme ruler of Mount Olympus]: but this is only what happens to mortals when one of us dies. As soon as the spirit leaves the white bones, the sinews no longer hold flesh and bones together—the blazing fire consumes them all; but the soul flits away fluttering like a dream. Make haste back to the light; but do not forget all this, tell it to your wife by and by.'

"As we were talking together, a crowd of women came up sent by awful Persephoneia, wives and daughters of great men. They gathered about the red blood, and I wondered how I should question them. This seemed to be the best plan. I drew my sword and kept off the crowd of ghosts; and then I let them form in a long line and come up one by one. Each one declared her lineage, then I questioned them all. . . .

[Odysseus resists the temptation to stay in Hadês with his mother and the other souls.]

". . . Then I went back at once to the ship, and told my men to loose the hawser [i.e., cable for mooring or towing a ship] and get away."

3. To Be Met with Acceptance

The earliest Buddhist scriptures in the Mahayana tradition of northern Asia are known as the *Tripitaka* (meaning "three baskets") and include the *Vinaya Pitaka*, the *Sutta Pitaka*, and the *Abhidhamma Pitaka*. The Mahayana tradition of Buddhism emphasizes the metaphysics of the thinking and the role of tradition and ritual. The *Sutta Pitaka* is thought to have been compiled around the first century CE. Here, *sutta* (or *sutra*) refers to a collection of aphorisms, or wise sayings, in the form of a manual (Bilhartz 2006, 210). In a selection entitled "Salla Sutta: The Arrow," found in the *Sutta Nipata*, the fifth book of the *Khuddaka Nikaya* in the *Sutta Pitaka*, the following is said about death: "Having reached old age, there is death. This is the natural course for a living being. With ripe fruits there is the constant danger that they will fall. In the same way, for those born and subject to death, there is always the fear of dying. Just as the pots made by a potter all end by being broken, so death is (the breaking up) of life" ("Salla Sutta: The Arrow" [1983] 1994 ¶1). The tone of acceptance of death as a natural phenomenon evidenced in this passage is found as well in the following piece of advice: "In this manner the world is afflicted by death and decay. But the wise do not grieve, having realized the nature of the world" ("Salla Sutta: The Arrow" [1983] 1994 ¶3). What follows is a selection of aphorisms or wise sayings on the acceptance of death from the *Sutta Nipata*.

Reading: "Salla Sutta: The Arrow," in the *Sutta Nipata* ([1983] 1994 [3.8]). Copyright © 1983 Buddhist Publication Society. Reproduced with the permission of the Buddhist Publication Society.

[*Sutta Nipata* 3.8]
Unindicated and unknown is the length of life of those subject to death. Life is difficult and brief and bound up with suffering. There is no means by which those who are born will not die. Having reached old age, there is death. This is the natural course for a living being. With ripe fruits there is the constant danger that they will fall. In the same way, for those born and subject to death, there is always the fear of dying. Just as the pots made by a potter all end by being broken, so death is (the breaking up) of life.

The young and old, the foolish and the wise, all are stopped short by the power of death, all finally end in death. Of those overcome by death and passing to another world, a father cannot hold back his son, nor relatives a relation. See! While the relatives are looking on and weeping, one by one each mortal is led away like an ox to slaughter.

In this manner the world is afflicted by death and decay. But the wise do not grieve, having realized the nature of the world. You do not know the path by which they came or departed. Not seeing either end you lament in vain. If any benefit is gained by lamenting, the wise would do it. Only a fool would harm himself. Yet through weeping and sorrowing the mind does not become calm, but still more suffering is produced, the body is harmed and one becomes lean and pale, one merely hurts oneself. One cannot protect a departed one (*peta*) by that means. To grieve is in vain.

By not abandoning sorrow a being simply undergoes more suffering. Bewailing the dead he comes under the sway of sorrow. See other men faring according to their deeds! Hence beings tremble here with fear when they come into the power of death. Whatever they imagine, it (turns out) quite different from that. This is the sort of disappointment that exists. Look at the nature of the world! If a man lives for a hundred years, or even more, finally, he is separated from his circle of relatives and gives up his life in the end. Therefore, having listened to the arahant [the Perfect One, the Buddha], one should give up lamenting. Seeing a dead body, one should know, "He will not be met by me again." As the fire in a burning house is extinguished with water, so a wise, discriminating, learned and sensible man should quickly drive away the sorrow that arises, as the wind (blows off) a piece of cotton. He who seeks happiness should withdraw the arrow: his own lamentations, longings and grief.

With the arrow withdrawn, unattached, he would attain to peace of mind; and when all sorrow has been transcended he is sorrow-free and has realized Nibbana [i.e., Nirvana, freedom from endless cycles of personal reincarnations with their consequent suffering].

4. To Be Met Calmly

The story of Gotami-tissa is one of the more famous ones in Buddhism in the Theravada tradition of South Asia. The Theravada tradition of Buddhism minimizes metaphysical thought and rituals and emphasizes practice centered in meditation. As the story goes, Gotami-tissa lost a child and experienced "sorrow-to-the-point-of-madness" (Olendzki 2013). She began to seek the help of others in order to understand this tragic occurrence. A neighbor encouraged her to seek the advice of the Buddha. In response, she brought the body of her dead son to him, asking him to cure him. Buddha instructed Gotami-tissa to go into the village and ask any villager who had not experienced death for a mustard seed, a household item that was commonly found in homes. Gotami went from house to house and could find no one who had not experienced death. She gathered many stories about death and mortality from the villagers. Through such stories, she began to realize that no human is free from a mortal existence. She returned to the Buddha and shared what she had learned:

> *It's not just a truth for one village or town,*
> *Nor is it a truth for a single family.*
> *But for every world settled by gods [and men]*
> *This indeed is what is true—impermanence.*
>   (Olendzki 2013, ¶8).

Gotami was awakened to a new understanding about life, and her fears of death were calmed.

The following version of her story is translated by Andrew Olendzki, who is faculty at the Barre Center for Buddhist Studies in Barre, Massachusetts.

Reading: "Skinny Gotami and the Mustard Seed," in *Access to Insight: Readings in Theravāda Buddhism* (Olendzki 2013). Reproduced with the permission of Andrew Olendzki.

After flowing-on for a hundred thousand ages,
she evolved in this Buddha-era among gods and men
in a poor family in Savatthi.
Her name was Gotami-tissa,
but because her body was very skinny
she was called "Skinny Gotami."
When she went to her husband's family,
she was scorned [and called] "daughter of a poor family."

Then she gave birth to a son,
and with the arrival of the son she was treated with respect.
But that son, running back and forth
and running all around, while playing met his end.
Because of this, sorrow-to-the-point-of-madness arose in her.
She thought: "Before I was one who received only scorn,
but starting from the time of the birth of my son I gained honor.
These [relatives] will now try to take my son,
in order to expose him outside [in the charnel ground]."

Under the influence of her sorrow-to-the-point-of-madness,
she took the dead corpse on her hip and
wandered in the city from the door of one house to another
[pleading]: "Give medicine to me for my son!"
People reviled her, [saying,] "What good is medicine?"
She did not grasp what they were saying.

And then a certain wise man, thinking
"This woman has had her mind deranged by sorrow for her son;
the ten-powered [Buddha] will know the medicine for her,"
said: "Mother, having approached the fully awakened one,
ask about medicine for your son."

She went to the vihara
at the time of the teaching of dhamma and said,

"Blessed One, give medicine to me for my son!"
The master, seeing her situation, said,
"Go, having entered the city,
into whatever house has never before experienced any death,
and take from them a mustard seed."

"Very well, Sir." [she replied],
and glad of mind she entered the city and came to the first house:
"The master has called for a mustard seed
in order to make medicine for my son.
If this house has never before experienced any death,
give me a mustard seed."
"Who is able to count how many have died here?"
"Then keep it. What use is that mustard seed to me?"
And going to a second and a third house,
her madness left her and her right mind was established
—thanks to the power of the Buddha.

She thought, "This is the way it will be in the entire city.
By means of the Blessed One's compassion for my welfare,
this will be what is seen."
And having gained a sense of spiritual urgency from that,
she went out and covered her son in the charnel ground.

She uttered this verse:
It's not just a truth for one village or town,
Nor is it a truth for a single family.
But for every world settled by gods [and men]
This indeed is what is true—impermanence.

And so saying, she went into the presence of the master.
Then the master said to her,
"Have you obtained, Gotami, the mustard seed?"
"Finished, sir, is the matter of the mustard seed" she said.
"You have indeed restored me."

And the master then uttered this verse:
A person with a mind that clings,

Deranged, to sons or possessions,
Is swept away by death that comes
— Like mighty flood to sleeping town.

At the conclusion of this verse, confirmed in the fruit of stream-entry,
she asked the master [for permission] to go forth [into the homeless life].
The master allowed her to go forth.
She gave homage to the master by bowing three times,
went to join the community of nuns,
and having gone forth, received her ordination.

It was not long before, through the doing of deeds with careful attention,
she caused her insight to grow . . . and she became an arahant.

5. To Be Grieved Only Outwardly

Greek philosopher and former Roman slave Epictetus (ca. 50–ca. 130 CE) was an exponent of Stoic ethics notable for its demand for self-management and personal freedom. In *The Handbook* (*The Enchiridion*), Epictetus shares his advice about living, including how to grieve the loss of another who has died. Here is his advice:

> When you see someone weeping in grief at the departure of his child or the loss of his property, take care not to be carried away by the appearance that the externals he is involved in are bad, and be ready to say immediately, "What weighs down on this man is not what has happened (since it does not weigh down on someone else), but his judgment about it." Do not hesitate, however, to sympathise with him verbally, and even to moan with him if the occasion arises; but be careful not to moan inwardly. (Epictetus 1983, 15–16)

While Epictetus's advice may seem harsh, the suggestion is this: when faced with death, feel free to participate in public displays of mourning, but be advised against privately grieving. Why grieve about something that one has no control over? Why become upset over something that one cannot change? Why worry about that which is out of one's hands? Instead, accept what is out of one's control and one will not be disappointed. One will not be disappointed because one's expectations will fit or be in line with what occurs. As Epictetus says, "Do not seek to have events happen as you want them to, but instead want them to happen as they do happen, and your life will go well" (1983, 13). What follows are some words of advice on death from an influential Stoic.

Reading: *The Handbook* (Epictetus 1983, 12, 13, 14, 15, 16, 19)

3. In the case of everything attractive or useful or that you are fond of, remember to say just what sort of thing it is, beginning with the least little things. If you are fond of a jug, say "I am fond of a jug!" For then when it is broken you will not be upset. If you kiss your child or your wife, say that you are kissing a human being; for when it dies you will not be upset.

5. What upsets people is not things themselves but their judgments about the things. For example, death is nothing dreadful (or else it would have appeared dreadful to Socrates), but instead the judgment about death that it is dreadful—*that* is what is dreadful. So when we are thwarted or upset or distressed, let us never blame someone else but rather ourselves, that is, our own judgments. An uneducated person accuses others when he is doing badly; a partly educated person accuses himself, an educated person accuses neither someone else nor himself.

8. Do not seek to have events happen as you want them to, but instead want them to happen as they do happen, and your life will go well.

9. Illness interferes with the body, not with one's faculty of choice [*proairesis*], unless that faculty of choice wishes it to. Lameness interferes with the limb, not with one's faculty of choice. Say this at each thing that happens to you, since you will find that it interferes with something else, not with you.

11. Never say of anything, "I have lost it," but instead, "I have given it back." Did your child die? It was given back. Did your wife die? She was given back. "My land was taken." So this too was given back. "But the person who took it was bad!" How does the way the giver [i.e., nature or god] asked for it back concern you? As long as he gives it, take care of it as something that is not your own, just as travelers treat an inn.

14. You are foolish if you want your children and your wife and your friends to live forever, since you are wanting things to be up to you that are not up to you, and things to be yours that are not yours. You are stupid in the same way if you want your slave boy to be faultless, since you are wanting badness not to be badness but something else. But wanting not to fail to get what you desire—*this* you are capable of. A person's master is someone who has power over what he wants or does not want, either to obtain it or take it away. Whoever wants to be free, therefore, let him not want or avoid anything that is up to others. Otherwise he will necessarily be a slave.

16. When you see someone weeping in grief at the departure of his child or the loss of his property, take care not to be carried away by the appearance that the externals he is involved in are bad, and be ready to say immediately, "What

weighs down on this man is not what has happened (since it does not weigh down on someone else), but his judgment about it." Do not hesitate, however, to sympathise with him verbally, and even to moan with him if the occasion arises; but be careful not to moan inwardly.

21. Let death and exile and everything that is terrible appear before your eyes every day, especially death; and you will never have anything contemptible in your thoughts or crave anything excessively.

27. Just as a target is not set up to be missed, in the same way nothing bad by nature happens in the world.

6. To Be Met with Tears and Pathos

C. S. Lewis (1898–1963) is probably most known for his children's book series *The Chronicles of Narnia*. In the series, children explore a realm beyond this life called Narnia, a fantasy world representing the good, as they encounter battles between good and evil. By profession, Lewis was a trained philosopher and Christian theologian at Oxford University in England. In his personal memoir *A Grief Observed*, Lewis recalls the death of his wife of many years and examines with detailed honesty the process of grief that he experienced. For Lewis, grief matters because death matters. As he says, "It is hard to have patience with people who say, 'There is no death' or 'Death doesn't matter.' There is death. And whatever is matters. And whatever happens has consequences, and it and they are irrevocable and irreversible" (1961, 16). In what follows, Lewis shows us how grief serves to keep the dead alive, even when this may bring about torment and confusion, bundles of questions, and unresolved grief. As he shares painstakingly regarding his deceased wife, "And her voice is still vivid. The remembered voice—that can turn me at any moment to a whimpering child" (1961, 16). In what follows, Lewis displays brutal honesty as he grieves the loss of his beloved wife and shows how death is met with tears and pathos. On this view, grief is the price one pays for love.

Reading: *A Grief Observed* (Lewis 1961, 7–9, 16). Copyright © C. S. Lewis Pte. Ltd. 1961. Extract reprinted by permission.

No one ever told me that grief felt so like fear. I am not afraid, but the sensation is like being afraid. The same fluttering in the stomach, the same restlessness, the yawning. I keep on swallowing.

At other times it feels like being mildly drunk, or concussed. There is a sort of invisible blanket between the world and me. I find it hard to take in what anyone says. Or perhaps, hard to want to take it in. It is so uninteresting. Yet I want the others to be about me. I dread the moments when the house is empty. If only they would talk to one another and not to me.

There are moments, most unexpectedly, when something inside me tries to assure me that I don't really mind so much, not so very much, after all. Love is not the whole of man's life. I was happy before I ever met H. [his deceased wife]. I've plenty of what are called "resources." People get over these things. Come, I shan't do so badly. One is ashamed to listen to this voice but it seems for a little to be making out a good case. Then comes a sudden jab of red-hot memory and all this "common sense" vanishes like an ant in the mouth of a furnace.

On the rebound one passes into tears and pathos. Maudlin tears. I almost prefer the moments of agony. These are at least clean and honest. But the bath of self-pity, the wallow, the loathsome sticky-sweet pleasure of indulging it—that disgusts me. And even while I'm doing it I know it leads me to misrepresent H. herself. Give that mood its head and in a few minutes I shall have substituted for the real woman a mere doll to be blubbered over. Thank God the memory of her is too strong (will it always be too strong?) to let me get away with it.

For H. [i.e., Helen Joy Davidman] wasn't like that at all. Her mind was lithe and quick and muscular as a leopard. Passion, tenderness, and pain were all equally unable to disarm it. It scented the first whiff of cant or slush; then sprang, and knocked you over before you knew what was happening. How many bubbles of mine she pricked! I soon learned not to talk rot to her unless I did it for the sheer pleasure—and there's another red-hot jab—of being exposed and laughed at. I was never less silly than as H.'s lover.

And no one ever told me about the laziness of grief. Except at my job—where the machine seems to run as much as usual—I loathe the slightest effort. Not only writing but even reading a letter is too much. Even shaving. What does it matter now whether my cheek is rough or smooth? They say an unhappy man wants distractions—something to take him out of himself. Only as a dog-tired man wants an extra blanket on a cold night; he'd rather lie there shivering than get up and find one. It's easy to see why the lonely become untidy; finally, dirty and disgusting.

Meanwhile, where is God? This is one of the most disquieting symptoms. When you are happy, so happy that you have no sense of needing Him, so happy that you are tempted to feel His claims upon you as an interruption, if you remember yourself and turn to Him with gratitude and praise, you will be—or so it feels—welcomed with open arms. But go to Him when your need is desperate, when all other help is vain, and what do you find? A door slammed in your face, and a sound of bolting and double bolting on the inside. After that, silence. You may as well turn away. The longer you wait, the more emphatic the silence will become. There are no lights in the windows. It might be an empty house. Was it ever inhabited? It seemed so once. And that seeming was as strong as this. What can this mean? Why is He so present a commander in our time of prosperity and so very absent a help in time of trouble?

I tried to put some of these thoughts to C. this afternoon. He reminded me that the same thing seems to have happened to Christ: "Why hast thou forsaken me?" I know. Does that make it easier to understand? . . .

It is hard to have patience with people who say, "There is no death" or "Death doesn't matter." There is death. And whatever is matters. And whatever happens has consequences, and it and they are irrevocable and irreversible. You might as well say that birth doesn't matter. I look up at the night sky. Is anything more certain than that in all those vast times and spaces, if I were allowed to search them, I should nowhere find her face, her voice, her touch? She died. She is dead. Is the word so difficult to learn?

I have no photograph of her that's any good. I cannot even see her face distinctly in my imagination. Yet the odd face of some stranger seen in a crowd this morning may come before me in vivid perfection the moment I close my eyes tonight. No doubt, the explanation is simple enough. We have seen the faces of those we know best so variously, from so many angles, in so many lights, with so many expressions—waking, sleeping, laughing, crying, eating, talking, thinking— that all the impressions crowd into our memory together and cancel out into a mere blur. But her voice is still vivid. The remembered voice—that can turn me at any moment to a whimpering child.

7. To Be Met with Denial, Anger, Bargaining, Depression, and Acceptance

Physician Elisabeth Kübler-Ross (1926–2004) was a Swiss-born psychiatrist and pioneer in near-death studies. Her work with dying patients occurred well before others devoted attention to this topic and resulted in the groundbreaking book *On Death and Dying* (1969). In the book, Kübler-Ross shares her observations of her time in conversation with dying patients. According to Kübler-Ross, talking about death and dying is important: "It might be helpful if more people would talk about death and dying as an intrinsic part of life just as they do not hesitate to mention when someone is expecting a baby. If this were done more often, we would not have to ask ourselves if we ought to bring this topic up with a patient, or if we should wait for the last admission [to the hospital]" (Kübler-Ross 1969, 150). In listening to patients talk about their dying experience, Kübler-Ross noticed some striking similarities in how they came to terms with their pending deaths. She summarized these similarities in terms of the "five stages of grief," namely, denial, anger, bargaining, depression, and acceptance. In the years following the publication of her work, the five stages of grief have been adopted by many physicians, nurses, therapists, and spiritual advisers (Elisabeth Kübler-Ross Foundation 2015). A significant body of literature and research on death and dying has developed as a result of Kübler-Ross's work on the ways humans grieve in the face of their own pending death and the death of another (Bonanno 2009; Konigsberg 2011). What follows is a series of excerpts on Kübler-Ross's account of the five stages of grief.

Reading: *On Death and Dying* (Kübler-Ross 1969, 49, 51, 52, 53, 67, 93, 95, 97, 98, 123–24, 147–48, 150). Reprinted with the permission of Scribner, a division of Simon and Schuster, Inc., from *On Death and Dying* by Dr. Elisabeth Kübler-Ross; copyright renewed © 1997 by Elisabeth Kübler-Ross. All rights reserved.

In the following pages is an attempt to summarize what we have learned from our dying patients in terms of coping mechanisms at the time of a terminal illness.

Among the over two hundred dying patients we have interviewed, most reacted to the awareness of a terminal illness at first with the statement, "No, not me, it cannot be true." This *initial* denial was as true for those patients who were told outright at the beginning of their illness as it was true for those who were not told explicitly and who came to this conclusion on their own a bit later on.

. . . Denial functions as a buffer after unexpected shocking news, allows the patient to collect himself and, with time, mobilize other, less radical defenses.

. . . Denial is usually a temporary defense and will soon be replaced by partial acceptance.

When the first stage of denial cannot be maintained any longer, it is replaced by feelings of anger, rage, envy, and resentment. The logical next question becomes, "Why me?" . . .

In contrast to the stage of denial, this stage of anger is very difficult to cope with from the point of view of family and staff. The reason for this is the fact that this anger is displaced in all directions and projected onto the environment at times almost at random. . . . The doctors are just no good, they don't know what tests to require and what diet to prescribe. . . . The nurses are even more often a target of their anger. Whatever they touch is not right. . . .

The third stage, the stage of bargaining, is less well known but equally help-ful to the patient, though only for brief periods of time. If we have been unable to face the sad facts in the first period and have been angry at people and God in the second phase, maybe we can succeed in entering into some sort of an agreement which may postpone the inevitable happening: "If God has decided to take us from this earth and he did not respond to my angry pleas, he may be more favorable if I ask nicely." . . .

The bargaining is really an attempt to postpone; it has to include a prize of-fered "for good behavior," it also sets a self-imposed "deadline" (e.g., one more performance, the son's wedding), and it includes an implicit promise that the pa-tient will not ask for more if this one postponement is granted. . . .

When the terminally ill patient can no longer deny his illness, when he is forced to undergo more surgery or hospitalization, when he begins to have more symptoms or becomes weaker and thinner, he cannot smile it off anymore. His numbness or stoicism, his anger and rage will soon be replaced with a sense of great loss [or depression].

. . . What we often tend to forget, however, is the preparatory grief that the terminally ill patient has to undergo in order to prepare himself for his final sepa-ration from this world. If I were to attempt to differentiate these two kinds of de-pressions, I would regard the first one a reactive depression, the second one a preparatory depression. . . .

If a patient has had enough time (i.e., not a sudden, unexpected death) and has been given some help in working through the previously described stages, he will reach a stage during which he is neither depressed nor angry about his "fate." He will have been able to express his previous feelings, his envy for the living and the healthy, his anger at those who do not have to face their end so soon. He will have mourned the impending loss of so many meaningful people and places and he will contemplate his coming end with a certain degree of quiet expectation. . . .

Acceptance should not be mistaken for a happy stage. It is almost devoid of feelings. It is as if pain has gone, the struggle is over. And there comes a time for "the final rest before the long journey" as one patient phrased it. . . .

We have discussed so far the different stages that people go through when they are faced with tragic news—defense mechanisms in psychiatric terms, cop-ing mechanisms to deal with extremely difficult situations. These means will last for different periods of time and will replace each other or exist at times side by side. The one thing that persists through all these stages is hope. . . .

It might be helpful if more people would talk about death and dying as an intrinsic part of life just as they do not hesitate to mention when someone is expecting a baby. If this were done more often, we would not have to ask ourselves if we ought to bring this topic up with a patient, or if we should wait for the last admission.

8. To Be Met with Culturally Inclusive Responses

Kelley Allen is a neonatal nurse practitioner at the Indiana University School of Medicine. Prior to this, she was the charge nurse for the neonatal transport team at Monroe Carell Jr. Children's Hospital at Vanderbilt University in Nashville, Tennessee. In this capacity, she was responsible for managing bedside care in the neonatal intensive care unit and for babies in need of medical care at outside facilities. While at Vanderbilt, she studied cultural and religious differences among grieving parents and families in the neonatal intensive care unit (NICU). She observed some striking similarities between and among traditions, namely, respect for the dying, the wishes of the family, and cherished religious or spiritual practices. She also observed significant similarities among members of particular ethnic groups. As she reports:

> The analysis of the research concerning the topic of culturally sensitive bereavement care formulated the emergence of several commonalities that are in direct contrast with the Westernized way of thinking about death and dying. In the Western culture, the idea of individualism and autonomy to make decisions is viewed as empowering and has come to be respected. . . . However, in the Eastern traditions of Hinduism, Judaism, Buddhism, and Muslim cultures, autonomy is seen as isolating and burdening upon the potentially sick or dying person. The school of thought for these cultures is generally that the family and/or community will make end-of-life decisions. (Allen 2013, 1)

The selection that follows highlights important cultural differences in the grief process. While readers may not agree with Allen's generalizations about particular cultural responses to death, the point of the passage is to appreciate differences among how people from different cultures grieve and how we can be more responsive practically to different ways of grieving.

Reading: "Eastern Traditions versus Western Beliefs as Related to the Grief Process Regarding End-of-Life Care in the Neonatal Intensive Care Unit" (Allen 2013)

The analysis of the research concerning the topic of culturally sensitive bereavement care formulated the emergence of several commonalities that are in direct contrast with the Westernized way of thinking about death and dying. In the Western culture, the idea of individualism and autonomy to make decisions is viewed as empowering and has come to be respected. This makes the idea of advanced directives, living wills, and do not resuscitate orders popular documents. However, in the Eastern traditions of Hinduism, Judaism, Buddhism, and Muslim cultures, autonomy is seen as isolating and burdening upon the potentially sick or dying person. The school of thought for these cultures is generally that the family and/or community will make end-of-life decisions. Therefore, there is no need for any type of advanced directives. In fact, sometimes just mentioning the idea of an advanced directive can create mistrust for the healthcare worker. This would include the prospect of withdrawal of care, which is generally considered taboo among these cultures for several reasons.

In Hinduism and Buddhism, there is strong belief in the idea of karma, destiny, and reincarnation. This means that nothing should be done to increase the process of dying because time of death has been predetermined by their God and would interfere with karma and successful reincarnation. The Jewish faith gives no merit to the Western concept of quality of life, but rather respects all life. They would, therefore, also do nothing to speed up the death of a loved one. Muslims generally view the idea of withdrawal of support equated with taking that person's soul. Since it is the family's priority to see that person die with a pure soul, withdrawal of care most likely will not be done. Muslims, Hindus, and Buddhists are opposed to comfort care measures because it interferes with karma or the repentance of sins. All of these cultures, including the Westernized cultures, have specific death rituals to be performed. All have a great desire for these practices to be respected and followed as closely as possible even when death takes place in a hospital setting. All Eastern cultures discussed value that the washing and preparation of the body be done by a same sex person of their same beliefs. This is an important factor to take into consideration if an infant of another culture dies suddenly in the NICU, since it is common practice for the bedside nurse to perform bathing and dressing of the body.

The Hindu, Buddhist, and Jewish cultures all expressed desires to have a designee remain with the deceased for extended time periods after death. The Muslim community, however, do not prefer to see nor stay with the patient after death. They also do not wish to receive any type of memento from the deceased, because all of these things tend to worsen their grief experience. This is important for healthcare workers to understand, because it is a vast change in bereavement practices in the NICU and could potentially cause complicated grief. Several of these Eastern cultures expressed the desire to pass away at home, but this is often not possible in the NICU setting.

Both Eastern and Western cultures are appreciative of early, open communication among the healthcare team, patient, family, and sometimes community leaders. These differing cultures also shared the need to repent and confess their sins to their religious leader of choice on their deathbed. This is commonly seen in Western society as well. Generally, the Eastern cultures were opposed to the process of an autopsy and desired to have all medical devices removed from their baby. This can sometimes cause conflict between the physician in the NICU and the family members depending upon the cause of death. Most of them had strong beliefs that burial or cremation should take place within twenty-four hours of death to prevent complications in the afterlife, no matter what the preferred disposition for the body.

Another common theme among these Eastern societies was long periods of mourning involving the entire family and community, which is a direct contrast to Western ideas of bereavement. In the Buddhist and Muslim communities, there are observed male and female role differences regarding end-of-life decision making and grieving processes. These variances can lead to issues with successful grieving due to the fact that men and women experience differing degrees of grief. Based on the discovered body of knowledge, findings note that no matter the culture, race, or religious beliefs, there are several common desires for the bereavement process. Respect for the dying, the wishes of the family, and their religious practices were found to be of utmost importance to many of the cultures studied. The need to receive information regarding diagnosis and outcome was deemed paramount to make informed decisions throughout the cultures surveyed. All of the cultures researched go through the grief process in some fashion. They each expressed the need for support from healthcare workers, friends, family, and their religious community leaders. Overall, there are many ways in which healthcare providers can assist in creating a culturally sensitive bereavement environment. This process may take more time and effort to discern the wishes of the family due to cultural barriers present.

Healthcare workers should never assume that all cultures can be treated the same.

To foster healthy grieving and to provide proper support for bereaving families, every NICU should create a bereavement program. It is apparent from the synthesis of this research that some type of culturally sensitive training should be a part of bereavement training. Interpreters should be readily available for these instances. There is also a need to have easier access to various religious leaders. In the field of nursing, each new graduate should be oriented to the bereavement process, and they should utilize a check list to ease their own discomfort and become more familiar with the process.

C. REFLECTIONS: GRIEF ASSESSMENT

1.(*) Epictetus tells us, "You are foolish if you want your children and your wife and your friends to live forever, since you are wanting things to be up to you that are not up to you, and things to be yours that are not yours" (1983, 15). Are we "foolish" if we want our loved ones to live forever? Are we "foolish" if we grieve the loss of our loved one? What do you think about what Epictetus tells us? Be sure to provide details in your reflections.

2.(*) Kelley Allen shares with us her experience in the clinical setting of culturally different ways of grieving. Select a culture different from your own and research how members of the culture typically grieve. Summarize your findings and give examples to illustrate your findings.

3.   Grief is a response to loss. Reflect briefly on the kinds of losses that have happened to you.
   a. Complete the list on the following page by checking the options in the columns on the right:
   b. Given your responses above (3.a), what can you say about how you grieve in the face of loss? What activities do you engage in when you grieve? What helps with your grieving? What complicates your grieving? Be sure to give details in your reflections and draw on your readings when appropriate.

D. FURTHER READINGS

Bonanno, George A. 2009. *The Other Side of Sadness: What the New Science of Bereavement Tells Us about Life after Loss.* New York: Basic Books.

Callanan, Maggie, and Patricia Kelley. [1992] 2012. *Final Gifts: Understanding the Special Awareness, Needs, and Communications of the Dying.* New York: Simon and Schuster.

Chidester, David. 2002. *Patterns of Transcendence: Religion, Death, and Dying.* Belmont, CA: Wadsworth.

Dixon, R. M. W. 1996. "Origin Legends and Linguistic Relationships." *Oceania* 67 (2): 127–40.

Doka, Kenneth J., and Terry L. Martin. 2010. *Grieving beyond Gender: Understanding the Ways Men and Women Mourn.* New York: Taylor and Francis.

	This hurt.	This did not hurt.	This has not happened to me.

Kinds of losses

1. loss of a family member (specify: _____)

2. loss of love (specify: _____)

3. loss of a friend (specify: _____)

4. loss of a pet (specify: _____)

5. loss of a favorite possession (specify:_____)

6. loss of a reputation (specify: _____)

7. loss of an ability (specify: _____)

8. loss of acceptance (specify: _____)

9. loss of place (specify:_____)

10. loss of spirituality (specify:_____)

11. loss of hope (specify:_____)

12. loss of trust (specify: _____)

13. loss of future possibility (specify: _____)

14. loss of time (specify: _____)

15. loss through theft (specify: _____)

16. loss of safety (specify: _____)

17. loss of body part (specify: _____)

18. loss of mind (specify: _____)

19. other (specify: _____)

20. other (specify: _____)

Kessler, David. 2007. *The Needs of the Dying: A Guide for Bringing Hope, Comfort, and Love to Life's Final Chapter*. New York: Harper Perennial.

Konigsberg, Ruth Davis. 2011. *The Truth about Grief: The Myth of Its Five Stages and the New Science of Loss*. New York: Simon and Schuster.

Kumar, Sameet M. 2005. *Grieving Mindfully: Compassionate and Spiritual Guide to Coping with Loss*. Oakland, CA: New Harbinger Publications.

Rando, Theresa A. 1984. *Grief, Dying, and Death: Clinical Interventions for Caregivers*. Champaign, IL: Research Press.

Schneider, John M. 2001. *Grief/Depression Assessment Inventory*. www.integraonline.org/assessments/grief_depression_inventory.pdf.

Part III

THE CHOICE
OF DEATH

In part III, death can be understood as that which can be hastened in the case of suicide, treatment refusal, or physician-assisted suicide. These discussions continue those that took place in part II in that they focus on the evaluations we make of death. In this way, discussions about death involve appeals to values or the ways we assign significance to life experiences. More specifically, they appeal to moral or ethical values concerning patient autonomy, nonmaleficence, beneficence, and justice (Beauchamp and Childress 2013). As a reminder, autonomy is the principle of respect for choice, nonmaleficence the principle of doing no harm, beneficence the principle of advancing the best interest of another, and justice the fair allocation of resources.

These moral principles arise in discussions about the choice to die. For instance, in discussions of death through suicide, a duty to advance the best interest of another supports the view that individuals should not be able to choose suicide because death through suicide is not in their best interest in that it undermines life and that others have a duty to prevent against such harm. Alternatively, in the case of suicide, a duty to advance the best interest of another who may be suffering terribly may support the view that death through suicide is praiseworthy. In discussions about death through treatment refusal, a duty of autonomy typically supports the view that competent adults should be able to refuse treatment at the end of life. Alternatively, in the case of treatment refusal, a sanctity-of-life standpoint typically supports the view that treatment refusal is never morally permissible because it undermines what is essential about life. In discussions about death through physician-assisted suicide, a duty of autonomy typically supports the view that competent adults should be able to request assistance in dying at the end of life. Alternatively, in the case of physician-assisted suicide, a duty to justice may support the view that society ought not discriminate against its most vulnerable members and that physician-assisted suicide is therefore considered unethical.

The readings in this third part highlight a range of ethical views of death that command our attention today. These include death as that which is hastened through suicide, hastened through treatment refusal, and hastened through physician-assisted suicide. These topics are presented through the lens of the nature of death, thus linking popular clinical and sociopolitical topics on death with discussions of the nature of death. In reading the following selections, look for ways in which we judge death to be good or bad, moral or immoral, and ethical and nonethical and what these judgments say about how we evaluate events involving death and dying. Again, readers are encouraged to take note of points of agreement and disagreement with authors in preparation for the exercises that appear at the end of each chapter.

— Chapter 12 —

To Be Hastened or Not

The Case of Suicide

A. CONTEXT

In *The Myth of Sisyphus*, French philosopher Albert Camus said, "There is only one really serious philosophical question, and that is suicide" (1955, 3). Why, for instance, would someone deprive him- or herself of what is necessary to live? What brings someone to a point in his or her life to choose against life? Do the moral mandates against suicide dating back to ancient times serve as any prevention against the decision to take one's life? In the early twenty-first century, much attention has been given to the topic of whether it is permissible to bring death about through the means of suicide. The discussion emerges in part because, as mentioned in chapter 2, A, death by suicide in the United States in 2009 was one of the top ten leading causes of death, leading to approximately 36,909 deaths (Heron 2012, 9). Even when one might hold that certain types of suicides may be permissible, one might still ask about what can and should be done about the number of suicides that occur.

What follows is a discussion of death from the perspective of whether it is permissible to bring death about through the act of suicide. On the one hand, the Roman philosopher Seneca holds that suicide is rational under certain conditions. The Japanese samurai Yamamoto Tsunetomo views suicide under certain conditions as an honorable act. The Scottish philosopher David Hume argues that suicide is permissible on grounds that it violates neither God's will nor one's duty to one's neighbors and society. Confucian philosopher Mencius argues that suicide is morally permissible if done for the sake of loyalty, self-sacrifice, and honor. On the other hand, Italian theologian and philosopher Thomas Aquinas argues that suicide is never permissible because it is contrary to an inclination of nature, harms the community, and is a sin against God. German philosopher Immanuel Kant argues that suicide is never permissible on grounds that it violates respect for persons. The US Catholic Bishops' *Catechism of the Catholic Church* echoes Aquinas's view that suicide is morally impermissible. The selections that follow illustrate a spectrum of views on the ethics of suicide and particular understandings of death that are valued and disvalued.

B. PERSPECTIVES

1. To Be Caused

a. Dependent on Quality of Life

Lucius Annaeus Seneca (or simply Seneca) (ca. 4 BCE–65 CE) was a Roman Stoic philosopher, statesman, and dramatist of Latin literature. He was tutor and later adviser to Emperor Nero (37–68 CE). His main ethical writings are contained in *Epistolae Morales* ("Moral Letters"). As a reminder, Stoic thinking embraces an ethic of consolation that consists of identifying with the impartial, inevitable order of the universe. Acceptance of such order brings peace of mind. Seneca wrote about suicide because he personally confronted it. As the story goes, Seneca was ordered by Nero, the last of the Julio-Claudian emperors (of which there were five, namely, Augustus, Tiberius, Caligula, Claudius, and Nero), to commit suicide for allegedly participating in a conspiracy to assassinate the emperor ("Seneca the Younger" 2017). In a letter to his friend Luciliuc, Seneca reflects on the act of suicide: "Dying early or late is of no relevance, dying well or ill is" (1987, 202). Following Nero's instruction, Seneca went to his death by poisoning himself and severing several of his veins.

Reading: "Letter to Luciliuc" (Seneca 1987, 202)

Living is not the good, but living well. The wise man therefore lives as long as he should, not as long as he can. . . . He will always think of life in terms of quality not quantity. . . . Dying early or late is of no relevance, dying well or ill is. . . . Even if it is true [that while there is life, there is hope], life is not to be bought at all costs.

b. Not Shameful

Yamamoto Tsunetomo (1659–1719) was a Japanese samurai or warrior who became a Buddhist monk in 1700 after the Shogunate government prohibited the practice of honor suicide. While in seclusion as a monk, Tsunetomo dictated his views about martial honor or valor to a younger samurai. The result is a handbook for the samurai, called *Hagakure: The Book of the Samurai*. The *Hagakure* consists of a series of short reflections on the philosophy of the true warrior. A true warrior is one who is willing and prepared to die for his master, by his own hands or the hands of another, and at any time. As Tsunetomo says, "The Way of the Samurai is found in death" (1979, 17). It is found in death because death marks a decisive moment where there is nothing else to do but die. At that moment, one faces an ultimate decision about who one is in life. The decision takes place without distraction or delay. Taken as a whole, the *Hagakure* is seen to be a handbook on the Way of Dying, since, as the following passage declares, the Way of the Samurai (*bushido* or samurai code of honor) is found in death.

Reading: *Hagakure: The Book of the Samurai* (Tsunetomo 1979, 17–18, 44, 45). © 1979, 2002 by William Scott Wilson. Reprinted by arrangement with The Permissions Company, Inc., on behalf of Shambhala Publications, Inc., Boulder, Colorado, www.shambhala.com.

[pp. 17–18]

Chapter 1

Although it stands to reason that a samurai should be mindful of the Way of the Samurai, it would seem that we are all negligent. Consequently, if someone were to ask, "What is the true meaning of the Way of the Samurai?" the person who would be able to answer promptly is rare. This is because it has not been established in one's mind beforehand. From this, one's unmindfulness of the Way can be known.

Negligence is an extreme thing.

The Way of the Samurai is found in death. When it comes to either/or, there is only the quick choice of death. It is not particularly difficult. Be determined and advance. To say that dying without reaching one's aim is to die a dog's death is the frivolous way of sophisticates. When pressed with the choice of life or death, it is not necessary to gain one's aim.

We all want to live. And in large part we make our logic according to what we like. But not having attained our aim and continuing to live is cowardice. This is a thin dangerous line. To die without gaining one's aim *is* a dog's death and fanaticism. But there is no shame in this. This is the substance of the Way of the Samurai. If by setting one's heart right every morning and evening, one is able to live as though his body were already dead, he gains freedom in the Way. His whole life will be without blame, and he will succeed in his calling. . . .

Calculating people are contemptible. The reason for this is that calculation deals with loss and gain, and the loss and gain mind never stops. Death is considered loss and life is considered gain. Thus, death is something that such a person does not care for, and he is contemptible. . . .

The saying of Shida Kichinosuke, "When there is a choice of either living or dying, as long as there remains nothing behind to blemish one's reputation, it is better to live," is a paradox. He also said, "When there is a choice of either going or not going, it is better not to go." A corollary to this would be, "When there is a choice of either eating or not eating, it is better not to eat. When there is a choice of either dying or not dying, it is better to die."

c. Freedom from Danger and Misery

Scottish philosopher David Hume (1711–76) is known for his contributions to epistemology and ethics. His lengthy *Treatise of Human Nature* is known for its incisive and provocative analysis of metaphysics and human knowledge. Hume is also known for his writings on ethics. During Hume's time, most people held that suicide was immoral on grounds that it violated God's will and one's duty to one's neighbors and society. Hume challenged these claims. In his writings, he argues that death through suicide may indeed be compatible with the will of a compassionate God. In addition, death through suicide may be in line with one's duties to one's neighbors and society. Death may actually be a good in that it can free us from the dangers and miseries of life. As Hume says in the selection that follows, "If it be no crime, both prudence and courage should engage us to rid ourselves at once of existence, when it becomes a burthen. It is the only way, that we can then be useful to society, by setting an example, which, if imitated, would preserve to every one his chance for happiness in life, and would effectually free him from all danger of misery" (1977, Su. 29).

In what follows, and well before others considered death through suicide to be a beneficial act, Hume argues in favor of suicide in cases in which more benefit than burden can be achieved through it. His view forecasts some of the positions one hears in contemporary discussions on suicide.

Reading: "Of Suicide" (Hume 1977, Su. 22–29)

Let us now examine, according to the method proposed, whether Suicide be of this kind of actions, and be a breach of our duty to our *neighbour* and to society.

A man, who retires from life, does no harm to society. He only ceases to do good; which, if it be an injury, is of the lowest kind. All our obligations to do good to society seem to imply something reciprocal. I receive the benefits of society, and therefore ought to promote its interest. But when I withdraw myself altogether from society, can I be bound any longer? But allowing, that our obligations to do good were perpetual, they have certainly some bounds. I am not obliged to do a small good to society, at the expence of a great harm to myself. Why then should I prolong a miserable existence, because of some frivolous advantage, which the public may, perhaps, receive from me? If upon account of age and infirmities, I may lawfully resign any office, and employ my time altogether in fencing against these calamities, and alleviating, as much as possible, the miseries of my future life: Why may I not cut short these miseries at once by an action, which is no more prejudicial to society? But suppose, that it is no longer in my power to promote the interest of the public: Suppose, that I am a burthen to it: Suppose, that my life hinders some person from being much more useful to the public. In such cases my resignation of life must not only be innocent but laudable. And most people, who lie under any temptation to abandon existence, are in some such situation. Those, who have health, or power, or authority, have commonly better reason to be in humour with the world.

A man is engaged in a conspiracy for the public interest; is seized upon suspicion; is threatened with the rack; and knows, from his own weakness, that the secret will be extorted from him: Could such a one consult the public interest better than by putting a quick period to a miserable life? This was the case of the famous and brave *Strozzi* of *Florence*. Again, suppose a malefactor justly condemned to a shameful death; can any reason be imagined, why he may not anticipate his punishment, and save himself all the anguish of thinking on its dreadful approaches? He invades the business of providence no more than the magistrate did, who ordered his execution; and his voluntary death is equally advantageous to society, by ridding it of a pernicious member.

That Suicide may often be consistent with interest and with our duty to *ourselves*, no one can question, who allows, that age, sickness, or misfortune may render life a burthen, and make it worse even than annihilation. I believe that no man ever threw away life, while it was worth keeping. For such is our natural horror of death, that small motives will never be able to reconcile us to it. And tho' perhaps the situation of a man's health or fortune did not seem to require this remedy, we may at least be assured, that any one, who, without apparent reason, has had recourse to it, was curst with such an incurable depravity or gloominess of temper, as must poison all enjoyment, and render him equally miserable as if he had been loaded with the most grievous misfortunes. If Suicide be supposed a crime, it is only cowardice that can impel us to it. If it be no crime, both prudence and courage should engage us to rid ourselves at once of existence, when it becomes a burthen. It is the only way, that we can then be useful to society, by setting an example, which, if imitated, would preserve to every one his chance for happiness in life, and would effectually free him from all danger of misery.

d. An Honor

Mencius (or "Mengzi" or "Meng Ke") was born around 372 BCE in the state of Zou, which today forms the territory of Zoucheng, Shandong Province, in China, eighteen miles south of Qufu, Confucius's birthplace. Mencius died around 289 BCE. He was an itinerant or traveling Chinese philosopher and sage, and one of the main interpreters of Confucianism. In his short passages and sayings, Mencius elaborates on the Confucian virtues or character traits of *ren* (benevolence, humaneness), *li* (propriety, observance of rites), *yi* (righteousness), and *zhi* (wisdom). He believes that human nature is naturally good and that part of being good is accepting nature's will without self-interested attachments. His view is evident in his account of death. As Mencius says, "I like life indeed, but there is that which I like more than life, and therefore, I will not seek to possess it by any improper ways. I dislike death indeed, but there is that which I dislike more than death, and therefore there are occasions when I will not avoid danger" ([1895] 2009, 2). For Mencius, dishonor is disliked more than death, so a dishonorable death is to be avoided and an honorable death is to be pursued. When an honorable death is made possible by suicide, suicide can be considered an honorable act.

Reading: *The Works of Mencius* (Mencius [1895] 2009 [6.1.10.1–8])

That it is proper to man's nature to love righteousness more than life, and how it is that many act as if it were not so.

1. Mencius said, "I like fish, and I also like bear's paws. If I cannot have the two together, I will let the fish go, and take the bear's paws. So, I like life, and I also like righteousness. If I cannot keep the two together, I will let life go, and choose righteousness.

2. "I like life indeed, but there is that which I like more than life, and therefore, I will not seek to possess it by any improper ways. I dislike death indeed, but there is that which I dislike more than death, and therefore there are occasions when I will not avoid danger.

3. "If among the things which man likes there were nothing which he liked more than life, why should he not use every means by which he could preserve it? If among the things which man dislikes there were nothing which he disliked more than death, why should he not do everything by which he could avoid danger?

4. "There are cases when men by a certain course might preserve life, and they do not employ it; when by certain things they might avoid danger, and they will not do them.

5. "Therefore, men have that which they like more than life, and that which they dislike more than death. They are not men of distinguished talents and virtue only who have this mental nature. All men have it; what belongs to such men is simply that they do not lose it.

6. "Here are a small basket of rice and a platter of soup, and the case is one in which the getting them will preserve life, and the want of them will be death;— if they are offered with an insulting voice, even a tramper will not receive them, or if you first tread upon them, even a beggar will not stoop to take them.

7. "And yet a man will accept of ten thousand chung, without any consideration of propriety or righteousness. What can the ten thousand chung add to him? When he takes them, is it not that he may obtain beautiful mansions, that he may secure the services of wives and concubines, or that the poor and needy of his acquaintance may be helped by him?

8. "In the former case the offered bounty was not received, though it would have saved from death, and now the emolument [compensation] is taken for the sake of beautiful mansions. The bounty that would have preserved from death was not received, and the emolument is taken to get the service of wives and concubines. The bounty that would have saved from death was not received, and the emolument is taken that one's poor and needy acquaintance may be helped by him. Was it then not possible likewise to decline this? This is a case of what is called—'Losing the proper nature of one's mind.'"

2. Not to Be Caused

a. Unlawful

Italian theologian and philosopher Thomas Aquinas (1225–74) was a Dominican priest and an influential scholar in the tradition of Scholasticism. Aquinas was the foremost classical proponent of natural theology, the study of the transcendent through reason and experience (as opposed to revelation). Aquinas held the view that the study of nature provides logical and empirical evidence that God exists and that the world is ordered in a providential way. Numerous treatises of modern philosophy in the West were conceived in developing or refuting his ideas, particularly in the areas of ethics, natural law, metaphysics, and political theory ("Thomas Aquinas" 2017). In the *Summa Theologica*, Aquinas raises the question "Whether it is lawful to kill oneself?" He argues that suicide is a sin for "three reasons" (1947 [II-II, q. 64, art. 5]): it is "contrary to an inclination in nature" to live, it "injures the community," and it is a sin "against God" the Creator. In keeping with a Scholastic formal writing format, Aquinas answers the question "Whether it is lawful to kill oneself?" by listing the objections to his answer to the question before sharing with us his answer. His view forecasts some of the positions one hears in contemporary discussions on suicide.

Reading: *Summa Theologica* (Aquinas 1947 [II-II, q. 64, art. 5])

Objection 1: It would seem lawful for a man to kill himself. For murder is a sin in so far as it is contrary to justice. But no man can do an injustice to himself, as is proved in Ethic. v, 11. Therefore no man sins by killing himself.

Objection 2: Further, it is lawful, for one who exercises public authority, to kill evil-doers. Now he who exercises public authority is sometimes an evil-doer. Therefore he may lawfully kill himself.

Objection 3: Further, it is lawful for a man to suffer spontaneously a lesser danger that he may avoid a greater: thus it is lawful for a man to cut off a decayed limb even from himself, that he may save his whole body. Now sometimes a man, by killing himself, avoids a greater evil, for example an unhappy life, or the shame of sin. Therefore a man may kill himself.

Objection 4: Further, Samson killed himself, as related in Judges 16, and yet he is numbered among the saints (Heb. 11). Therefore it is lawful for a man to kill himself.

Objection 5: Further, it is related (2 Mac. 14:42) that a certain Razias killed himself, "choosing to die nobly rather than to fall into the hands of the wicked, and to suffer abuses unbecoming his noble birth." Now nothing that is done nobly and bravely is unlawful. Therefore suicide is not unlawful.

On the contrary, Augustine says (De Civ. Dei i, 20): "Hence it follows that the words 'Thou shalt not kill' refer to the killing of a man—not another man; therefore, not even thyself. For he who kills himself, kills nothing else than a man."

I answer that, it is altogether unlawful to kill oneself, for three reasons. First, because everything naturally loves itself, the result being that everything naturally keeps itself in being, and resists corruptions so far as it can. Wherefore suicide is contrary to the inclination of nature, and to charity whereby every man should love himself. Hence suicide is always a mortal sin, as being contrary to the natural law and to charity. Secondly, because every part, as such, belongs to the whole. Now every man is part of the community, and so, as such, he belongs to the community. Hence by killing himself he injures the community, as the Philosopher declares (Ethic. v, 11). Thirdly, because life is God's gift to man, and is subject to His power, Who kills and makes to live. Hence whoever takes his own life, sins against God, even as he who kills another's slave, sins against that slave's master, and as he who usurps to himself judgment of a matter not entrusted to him. For it belongs to God alone to pronounce sentence of death and life, according to Dt. 32:39, "I will kill and I will make to live."

Reply to Objection 1: Murder is a sin, not only because it is contrary to justice, but also because it is opposed to charity which a man should have towards himself: in this respect suicide is a sin in relation to oneself. In relation to the community and to God, it is sinful, by reason also of its opposition to justice.

Reply to Objection 2: One who exercises public authority may lawfully put to death an evil-doer, since he can pass judgment on him. But no man is judge of himself. Wherefore it is not lawful for one who exercises public authority to put himself to death for any sin whatever: although he may lawfully commit himself to the judgment of others.

Reply to Objection 3: Man is made master of himself through his free-will: wherefore he can lawfully dispose of himself as to those matters which pertain to this life which is ruled by man's free-will. But the passage from this life to another and happier one is subject not to man's free-will but to the power of God. Hence it is not lawful for man to take his own life that he may pass to a happier life, nor that he may escape any unhappiness whatsoever of the present life, because the ultimate and most fearsome evil of this life is death, as the Philosopher states (Ethic. iii, 6). Therefore to bring death upon oneself in order to escape the other afflictions of this life, is to adopt a greater evil in order to avoid a lesser. In like manner it is unlawful to take one's own life on account of one's having committed a sin, both because by so doing one does oneself a very great injury, by depriving oneself of the time needful for repentance, and because it is not lawful to slay an evildoer except by the sentence of the public authority. Again it is unlawful for a woman to kill herself lest she be violated, because she ought not to commit on herself the very great sin of suicide, to avoid the lesser sin of another. For she commits no sin in being violated by force, provided she does not consent, since "without consent of the mind there is no stain on the body," as the Blessed Lucy declared. Now it is evident that fornication and adultery are less grievous sins than taking a man's, especially one's own, life: since the latter is most grievous, because one injures oneself, to whom one owes the greatest love. Moreover it is most dangerous since no time is left wherein to expiate it by repentance. Again it is not lawful for anyone to take his own life for fear he should consent to sin, because "evil must not be done that good may come" (Rom. 3:8) or that evil may be avoided especially if the evil be of small account and an uncertain event, for it is uncertain whether one will at some future time consent to a sin, since God is able to deliver man from sin under any temptation whatever.

Reply to Objection 4: As Augustine says (De Civ. Dei i, 21), "not even Samson is to be excused that he crushed himself together with his enemies under the ruins of the house, except the Holy Ghost, Who had wrought many wonders through him, had secretly commanded him to do this." He assigns the same reason in the case of certain holy women, who at the time of persecution took their own lives, and who are commemorated by the Church.

Reply to Objection 5: It belongs to fortitude that a man does not shrink from being slain by another, for the sake of the good of virtue, and that he may avoid sin. But that a man take his own life in order to avoid penal evils has indeed an appearance of fortitude (for which reason some, among whom was Razias, have killed themselves thinking to act from fortitude), yet it is not true fortitude, but rather a weakness of soul unable to bear penal evils, as the Philosopher (Ethic. iii, 7) and Augustine (De Civ. Dei 22, 23) declare.

b. Not to Be Chosen

German philosopher Immanuel Kant (1724–1804) was from Königsberg, Germany, which today is Kaliningrad, Russia. He researched, lectured, and wrote on philosophy and anthropology during a period called the Enlightenment at the end of the eighteenth century. Scholars of the Enlightenment emphasized the role reason plays in securing knowledge. Kant's major work, the *Critique of Pure Reason* (*Kritik der reinen Vernunft*), aimed to unite reason with experience to move beyond what he took to be failures of traditional philosophy and metaphysics. Kant hoped to end an age of speculation where objects outside experience were used to support what he saw as futile theories as provided by earlier metaphysicians such as Thomas Aquinas (see B.2.a in this chapter), while opposing the skepticism of thinkers such as David Hume (see B.1.c in this chapter) ("Immanuel Kant" 2017). In the *Groundwork of the Metaphysics of Morals* ([1785] 2012) (also known as *Grundlegung zur Metaphysik der Sitten, Foundations of the Metaphysics of Morals, Grounding of the Metaphysics of Morals*, and *Grounding for the Metaphysics of Morals*), Kant argues that suicide is a violation of the moral law or what he calls the "categorical imperative": "Now we see at once that a system of nature of which it should be a law to destroy life by means of the very feeling whose special nature it is to impel to the improvement of life would contradict itself and, therefore, could not exist as a system of nature; hence that maxim cannot possibly exist as a universal law of nature and, consequently, would be wholly inconsistent with the supreme principle of all duty" (Kant [1785] 2012, sec. 2). In other words, suicide is a violation of the categorical imperative because the action of ending one's life does not treat humanity as expressed in one's own person as an end in itself. Here, a person is an "end in itself" when he or she is considered an autonomous being worthy of dignity or self-worth. The problem with suicide is that it treats humanity as expressed in one's own person as a means to pleasure and the avoidance of pain. Because treating humanity as expressed in one's own person as a means to pleasure and the avoidance of pain is morally wrong, suicide is considered immoral. What follows is a selection from Kant that sets forth his view on the moral impermissibility of suicide.

Reading: *Groundwork of the Metaphysics of Morals* (Kant [1785] 2012, §2)

There is therefore but one categorical imperative, namely, this: Act only on that maxim whereby thou canst at the same time will that it should become a universal law. Now if all imperatives of duty can be deduced from this one imperative as from their principle, then, although it should remain undecided what is called duty is not merely a vain notion, yet at least we shall be able to show what we understand by it and what this notion means. Since the universality of the law according to which effects are produced constitutes what is properly called nature in the most general sense (as to form), that is the existence of things so far as it is determined by general laws, the imperative of duty may be expressed thus: Act as if the maxim of thy action were to become by thy will a universal law of nature. We will now enumerate a few duties, adopting the usual division of them into duties to ourselves and ourselves and to others, and into perfect and imperfect duties.

    1. A man reduced to despair by a series of misfortunes feels wearied of life, but is still so far in possession of his reason that he can ask himself whether it would not be contrary to his duty to himself to take his own life. Now he inquires whether the maxim of his action could become a universal law of nature. His

maxim is: "From self-love I adopt it as a principle to shorten my life when its longer duration is likely to bring more evil than satisfaction." It is asked then simply whether this principle founded on self-love can become a universal law of nature. Now we see at once that a system of nature of which it should be a law to destroy life by means of the very feeling whose special nature it is to impel to the improvement of life would contradict itself and, therefore, could not exist as a system of nature; hence that maxim cannot possibly exist as a universal law of nature and, consequently, would be wholly inconsistent with the supreme principle of all duty.

c. Contrary to Self, Others, and God

The *Catechism of the Catholic Church* is written by a group of cardinals and bishops (i.e., senior leaders in the church) and is a summary of the beliefs of members of the Roman Catholic faith. In 1985, at a synod of cardinals and bishops in Rome convened to celebrate the twentieth anniversary of the conclusion of the Second Vatican Council (1962–65), a proposal to develop a universal catechism for the Roman Catholic Church was made and accepted. The outcome was the *Catechism of the Catholic Church*, first published in 1992. A new edition with some modifications was released in 1997.

The *Catechism* addresses a wide range of issues and practices. With regard to suicide, it states that "suicide contradicts the natural inclination of the human being to preserve and perpetuate his life. It is gravely contrary to the just love of self. It likewise offends love of neighbor because it unjustly breaks the ties of solidarity with family, nation, and other human societies to which we continue to have obligations. Suicide is contrary to love for the living God" (*Catechism* 1997, 2281). What follows is Roman Catholic teaching on suicide. Cross-references to related church teachings in the *Catechism* are found in parentheses at the end of each section.

Reading: "Suicide," in *Catechism of the Catholic Church* (1997 [¶¶2280–83]). Copyright © 1997, United States Catholic Conference, Inc.—Libreria Editrice Vaticana.

2280 Everyone is responsible for his life before God who has given it to him. It is God who remains the sovereign Master of life. We are obliged to accept life gratefully and preserve it for his honor and the salvation of our souls. We are stewards, not owners, of the life God has entrusted to us. It is not ours to dispose of. (2258)

2281 Suicide contradicts the natural inclination of the human being to preserve and perpetuate his life. It is gravely contrary to the just love of self. It likewise offends love of neighbor because it unjustly breaks the ties of solidarity with family, nation, and other human societies to which we continue to have obligations. Suicide is contrary to love for the living God. (2212)

2282 If suicide is committed with the intention of setting an example, especially to the young, it also takes on the gravity of scandal. Voluntary co-operation in suicide is contrary to the moral law. (1735)

Grave psychological disturbances, anguish, or grave fear of hardship, suffering, or torture can diminish the responsibility of the one committing suicide.

2283 We should not despair of the eternal salvation of persons who have taken their own lives. By ways known to him alone, God can provide the opportunity for salutary repentance. The Church prays for persons who have taken their own lives. (1037)

C. REFLECTIONS: LAST WILL AND TESTAMENT

1. (*) Summarize the reasons that are given for why suicide is considered immoral and why it is considered moral. Draw on philosophical, clinical, and/or sacred text analysis in your summary and give details.

2. (*) Take a position: Do you think suicide is immoral or moral? Draw on philosophical, clinical, and/or sacred text analysis in your summary and give details.

3. This is a continuation of your advance planning in preparation for death. Your task is to complete your last will and testament. A last will and testament allows you to leave your wishes after you die, select an executor of your will (who will carry out your wishes), designate your beneficiaries (or recipients of your possessions), and assign guardians for minor children (if applicable). It also allows you to specify a financial power of attorney, i.e., a person who will make financial decisions for you when you become incapacitated or unable to make decisions on your own.

   Here's the task: Find a "last will and testament" (or "last will") in the state in which you reside online. Do not pay for one. Complete the form or follow the directions for writing a last will and testament. Notes: (i) This is a legal document in the state in which you reside. (ii) If you are serious about drafting a will, you are encouraged to seek the advice and assistance from an attorney who specializes in the area of estate planning in completing your last will and testament. State, regional, and national laws vary and need to be taken into account in drafting a will.

D. FURTHER READINGS

Barry, Vincent. 2007. *Philosophical Thinking about Death and Dying*. Belmont, CA: Thomson Wadsworth.

Camus, Albert. 1955. *The Myth of Sisyphus and Other Essays*. Translated by Justin O'Brien. New York: Alfred A. Knopf.

Cholbi, Michael. 2016. "Suicide." In *Stanford Encyclopedia of Philosophy*. http://plato.stanford .edu/entries/suicide/.

Durkheim, Emile. 1951. *On Suicide: A Study in Sociology*. New York: Free Press.

Heron, Melonie. 2012. "Deaths: Leading Causes for 2009." *National Vital Statistics Reports* 61 (7): 1–96.

Humphry, Derek. 2002. *Final Exit: The Practicalities of Self-Deliverance and Assisted Suicide for the Dying*. 3rd ed. New York: Dell.

Jaworski, Katrina. 2010. "The Gender-ing of Suicide." *Australian Feminist Studies* 25 (63): 47–61.

Schneidman, Edwin S. 2004. *Autopsy of a Suicidal Mind*. New York: Oxford University Press.

Seneca, Lucius Annaeus. 2010. *Ad Lucilium Epistulae Morales*. Vol. 1. N.p.: Nabu Press.

Chapter 13

To Be Hastened or Not

The Case of Treatment Refusal

A. CONTEXT

Today, medicine offers the ability to treat disease quite effectively, thereby prolonging the life of those who have access to medical interventions. When this prolongation is not in concert with a patient's choice or welfare, a patient might refuse
treatment. In the United States, treatment refusal in the medical setting is legally
supported by the US Patient Self-Determination Act of 1991, which recognizes a patient's right to refuse treatment based on his or her autonomy. Treatment refusal is
also supported by what are called "advance directives," legal documents that indicate
when life-prolonging treatment should be withdrawn or withheld, thereby allowing the disease process to take its natural course. A "living will" allows an adult patient eighteen years and older who is considered "terminal" to refuse treatment in
the event he or she becomes incompetent. A "medical durable power of attorney" or
"durable power of attorney for health care" allows an adult patient to designate another to speak on his or her own behalf and refuse lifesaving treatment when appropriate. Versions of advance directives are found in every state in the United States as
well as in the countries of Germany, Italy, Switzerland, and England ("Euthanasia
and Physician-Assisted Suicide" 2016).

In the United States, the POLST (Physician's Orders for Life-Sustaining Treatment) paradigm (National POLST Paradigm Office 2016) complements the use of
advance directives. The POLST Paradigm was developed to improve the quality of

patient care at the end of life by identifying patients' wishes regarding medical treatment. The POLST form allows patients to express particular wishes regarding treatment, including the refusal of treatment, at the end of life. A POLST form is found in most states.

What follows is a discussion of death from the perspective of whether it is permissible to refuse treatment knowing that death may soon occur. On the one hand, the Colorado Living Will and Colorado Medical Durable Power of Attorney for Health Care allow a patient to refuse clinical treatment. Readers might note that statutes for a living will and a durable power of attorney for health care are found in all states in the United States, and a reader may wish to look at the legislation found in his or her own state. Physician Leung Wing Chu discusses the emerging support of advance directives among nursing home residents in Hong Kong. On the other hand, physician and bioethicist Abraham Rudnick raises concern about the issue of treatment refusal in the psychiatric setting. Journalist Carol Forsloff raises a number of concerns about a parent refusing treatment on behalf of a child. Physicians Jung Kwak and William Haley discuss concerns about treatment refusal found among members of diverse racial and ethnic groups. The selections that follow illustrate a spectrum of views on the ethics of treatment refusal and particular understandings of death that are valued and disvalued.

B. PERSPECTIVES

1. To Be Accepted

a. My Right

With the rise of the development and use of high-technology medicine in the twentieth century, medicine has become increasingly successful in treating disease and illness. One of the side effects of such treatment is that patients live longer and, for some, with terminal illnesses. Some welcome the opportunity to live a few more years; others wish not to have such years. For those who do not wish to have their life prolonged, legislation has been developed that allows individual patients to choose to discontinue medical treatment, thereby allowing the natural disease process to take its course and end in death.

In the mid-1970s in the United States, states began to pass legislation recognizing the right of terminal patients to refuse treatment when they are no longer able to speak on their own behalf. California was the first state in the United States in 1975 to pass such legislation. Legislation outlines the requirements for a living will and, although legislation can vary from state to state, all provide the guidelines for completing such forms. An individual is to complete the form in the state in which he or she is considered a resident. Forms are available online and at local hospitals in every state. What follows is an example of legislation for a living will from the state of Colorado.

Reading: "Colorado Medical Treatment Decision Act" (2012, C.R.S. 15-18-104)

Title 15. Probate, Trusts, and Fiduciaries
Declarations—Future Medical Treatment
Article 18. Colorado Medical Treatment Decision Act (2012, C.R.S. 15-18-104).
Declaration as to medical treatment

(1) Any adult with decisional capacity may execute a declaration directing that life-sustaining procedures be withheld or withdrawn if, at some future time, he or she has a terminal condition or is in a persistent vegetative state, and lacks decisional capacity to accept or reject medical or surgical treatment. It shall be the responsibility of the declarant or someone acting for the declarant to provide the declaration to the attending physician or advanced practice nurse for entry in the declarant's medical record.

(2) In the case of a declaration of a qualified patient known to the attending physician to be pregnant, a medical evaluation shall be made as to whether the fetus is viable. If the fetus is viable, the declaration shall be given no force or effect until the patient is no longer pregnant.

(3) (a) A declaration may contain separate written statements regarding the declarant's preference concerning life-sustaining procedures and artificial nutrition and hydration if the declarant has a terminal condition or is in a persistent vegetative state.

(b) The declarant may provide in his or her declaration one of the following actions:

(I) That artificial nutrition and hydration not be continued;

(II) That artificial nutrition and hydration be continued for a specified period; or

(III) That artificial nutrition and hydration be continued.

(4) Notwithstanding the provisions of subsection (3) of this section and section 15-18-103 (10), when an attending physician or advanced practice nurse has determined that pain results from a discontinuance of artificial nutrition and hydration, the physician or advanced practice nurse may order that artificial

nutrition and hydration be continued to the extent necessary to provide comfort and alleviate pain.

(5) A declaration executed before two witnesses by any adult with decisional capacity shall be legally effective for the purposes of this article.

(6) A declaration executed pursuant to this article may include a document with a written statement as provided in section 12-34-105 (a), C.R.S., or a written statement in substantially similar form, indicating a decision regarding organ and tissue donation. Such a document shall be executed in accordance with the provisions of the "Revised Uniform Anatomical Gift Act," part 1 of article 34 of title 12, C.R.S.

(7) A declaration executed pursuant to this article may be combined with a medical power of attorney to create a single document. Such a document shall comply with all requirements of this title and in accordance with the provisions of the "Colorado Patient Autonomy Act," sections 15-14-503 to 15-14-509.

(8) A declaration executed pursuant to this article may include a written statement In which the declarant designates individuals with whom the declarant's attending physician, any other treating physician, or another medical professional may speak concerning the declarant's medical condition prior to a final determination as to the withholding or withdrawal of life-sustaining procedures, including artificial nutrition and hydration. The designation of such individuals in the document shall be considered to be consistent with the privacy requirements of the federal "Health Insurance Portability and Accountability Act of 1996," 42 U.S.C. sec. 1320d to 1320d-8, as amended, referred to in this section as "HIPAA," regarding waiver of confidentiality.

(9) A declaration executed pursuant to this article may include a written statement providing individual medical directives from the declarant to the attending physician or any other treating medical personnel.

b. My Right Voiced by My Agent

For some, the living will (see B.1.a in this chapter) does not give sufficient protection of a patient's end-of-life wishes. It does not because the living will applies only if a patient is diagnosed with a terminal condition. A patient has a terminal condition if he or she has six months to live. Typical terminal scenarios involve advanced stages of cancer, but not dementia or chronic clinical conditions. In addition, a living will does not give sufficient protection because it does not allow a patient to choose another to speak on his or her behalf in the event he or she is unable to do so. The durable power of attorney for health care (or as referred to in the Colorado legislation, the medical durable power of attorney) offers the opportunity to designate another to speak on a patient's behalf after the patient becomes unable to do so. In the mid-1980s in the United States, states began to pass legislation allowing an individual to choose an agent to make medical decisions in the event he or she is unable to make them. These decisions include refusing medical treatment. Such legislation protects agents from being accused of murder or manslaughter. Although legislation varies from state to state, all states in the United States provide the guidelines for completing such forms. Individuals are to complete the form in the state in which they are considered a resident. Forms are available online and at local hospitals in every state. What follows is an example of a medical durable power of attorney from the state of Colorado.

Reading: "Colorado Medical Durable Power of Attorney" (2012, Title 15, C.R.S. 15-14-506)

Title 15. Probate, Trusts, and Fiduciaries
Colorado Probate Code
Article 14. Persons under Disability—Protection
Part 5. Powers of Attorney C.R.S. 15-14-506 (2012)
15-14-506. Medical durable power of attorney

(1) The authority of an agent to act on behalf of the principal in consenting to or refusing medical treatment, including artificial nourishment and hydration, may be set forth in a medical durable power of attorney. A medical durable power of attorney may include any directive, condition, or limitation of an agent's authority.

(2) The agent shall act in accordance with the terms, directives, conditions, or limitations stated in the medical durable power of attorney, and in conformance with the principal's wishes that are known to the agent. If the medical durable power of attorney contains no directives, conditions, or limitations relating to the principal's medical condition, or if the principal's wishes are not otherwise known to the agent, the agent shall act in accordance with the best interests of the principal as determined by the agent.

(3) An agent appointed in a medical durable power of attorney may provide informed consent to or refusal of medical treatment on behalf of a principal who lacks decisional capacity and shall have the same power to make medical treatment decisions the principal would have if the principal did not lack such

decisional capacity. An agent appointed in a medical durable power of attorney shall be considered a designated representative of the patient and shall have the same rights of access to the principal's medical records as the principal. In making medical treatment decisions on behalf of the principal, and subject to the terms of the medical durable power of attorney, the agent shall confer with the principal's attending physician concerning the principal's medical condition.

(3.5) Any medical durable power of attorney executed under sections 15-14-503 to 15-14-509 may also have a document with a written statement as provided in section 12-34-105 (b), C.R.S., or a statement in substantially similar form, indicating a decision regarding organ and tissue donation. Such a document shall be executed in accordance with the provisions of the "Revised Uniform Anatomical Gift Act," part 1 of article 34 of title 12, C.R.S. Such a written statement may be in the following form:

I hereby make an anatomical gift, to be effective upon my death, of:

A. ____ Any needed organs/tIssues

B. ____ The following organs/tissues: _____

Donor signature: _____

(4) (a) Nothing in this section or in a medical durable power of attorney shall be construed to abrogate or limit any rights of the principal, including the right to revoke an agent's authority or the right to consent to or refuse any proposed medical treatment, and no agent may consent to or refuse medical treatment for a principal over the principal's objection.

(b) Nothing in this article shall be construed to supersede any provision of article 1 of title 25, C.R.S., or article 10.5 or article 65 of title 27, C.R.S.

(5) (a) Nothing in this part 5 shall have the effect of modifying or changing the standards of the practice of medicine or medical ethics or protocols.

(b) Nothing in this part 5 or in a medical durable power of attorney shall be construed to compel or authorize a health care provider or health care facility to administer medical treatment that is otherwise illegal, medically inappropriate, or contrary to any federal or state law.

(c) Unless otherwise expressly provided in the medical durable power of attorney under which the principal appointed the principal's spouse as the agent, a subsequent divorce, dissolution of marriage, annulment of marriage, or legal separation between the principal and spouse appointed as agent automatically revokes such appointment. However, nothing in this paragraph (c) shall be construed to revoke any remaining provisions of the medical durable power of attorney.

(d) Unless otherwise specified in the medical durable power of attorney, if a principal revokes the appointment of an agent or the agent is unable or unwilling to serve, the appointment of the agent shall be revoked. However, nothing in this paragraph (d) shall be construed to revoke any remaining provisions of the medical durable power of attorney.

(6) (a) This part 5 shall apply to any medical durable power of attorney executed on or after July 1, 1992. Nothing in this part 5 shall be construed to modify or affect the terms of any durable power of attorney executed before such date and which grants medical treatment authority. Any such previously executed durable power of attorney may be amended to conform to the provisions of this part 5. In the event of a conflict between a medical durable power of attorney executed pursuant to this part 5 and a previously executed durable power of attorney, the provisions of the medical durable power of attorney executed pursuant to this part 5 shall prevail.

(b) Unless otherwise specified in a medical durable power of attorney, nothing in this part 5 shall be construed to modify or affect the terms of a declaration executed in accordance with the "Colorado Medical Treatment Decision Act," article 18 of this title.

c. Preferred in Certain Situations

Advance directives are found in other countries besides the United States. Physician Leung Wing Chu is faculty in the Division of Geriatric Medicine, the Department of Medicine, and Queens Mary Hospital at the University of Hong Kong. He has been writing on advance directives and their use in Hong Kong for many years. In his work, Chu studies the use of advance directives by nursing home residents in Hong Kong. As he reports,

> In Hong Kong, most Chinese patients and the lay public are not familiar with advance directives. In a recent local study involving 1600 Chinese elderly persons living in old-age homes, Chu et al. reported that 96% of them had no knowledge about advance directives. However, after explanation of the meaning and value of an advance directive, 88% of them wanted to have their own advance directives. Most of them also agreed that advance directives can help express their preferences regarding future medical treatments and end-of-life care decisions, in case they go on to become mentally incompetent. (2012, 176)

What follows is an editorial appearing in the *Hong King Medical Journal* that addresses some of the challenges patients have with accepting treatment refusal directives in Chinese culture.

Reading: "One Step Forward for Advance Directives in Hong Kong" (Chu 2012, 176–77). Reproduced with permission from the *Hong Kong Medical Journal.* Hong Kong Academy of Medicine.

In Hong Kong, most Chinese patients and the lay public are not familiar with advance directives. In a recent local study involving 1600 Chinese elderly persons living in old-age homes, Chu et al. reported that 96% of them had no knowledge about advance directives. However, after explanation of the meaning and value of an advance directive, 88% of them wanted to have their own advance directives. Most of them also agreed that advance directives can help express their preferences regarding future medical treatments and end-of-life care decisions, in case they go on to become mentally incompetent. In another study of mostly healthy community-living Chinese adults, Pang et al. also reported that 77% of the subjects wanted to have their own advance directives. In a third local survey on elderly patients with chronic medical illnesses, 49% of them would use advance directives if there was suitable legislation in Hong Kong. These local studies show a general acceptance of the advance directive concept among our local Chinese population, and imply that it is time to implement their use in Hong Kong.

In this issue of the *Journal*, Wong et al. have successfully demonstrated the feasibility of implementing advance directives in their patients. In their prospective study of 191 patients with advanced cancers, 63% of the subjects opted to engage advance directives, which was in accord with the finding of 68% having advance directives in a recent US study. The most important factor associated with having a decision was insight about their end-stage malignancy illness. In line with this finding, Chu et al had previously reported that the strongest predictor associated with a preference for advance directives among Chinese elderly old-age home residents was the wish to be informed of a terminal illness. Therefore, the most important factor influencing engagement of an advance directive seems to be the patient's knowledge and insight about his or her terminal illness.

Another borderline factor is the attitude of the family members towards an advance directive. In this regard, the Chinese culture may affect the decisionmaking process in end-of-life care decisions. The Chinese often view overt reference to death as a taboo and would not like to talk about death. In relation to most end-of-life care decisions, Chinese family members often play a very influential role. Very often, patients prefer consulting their relatives before making health-care decisions. In Wong et al.'s study, objection by the family was a borderline factor, showing a trend against engaging in advance directives. However, this occurred uncommonly, as only 6% of those not engaging in advance directives did so because of family objections.

Taking the message of this study forward, it is time to further promulgate the advance directive concept in patients with terminal cancer and other end-stage diseases in all Hong Kong hospitals. To ensure a high success rate in helping patients to enact their advance directives, clinicians should provide detailed explanations about disease prognoses and palliative treatments to both the patients and their family. This could improve patient insights about their illnesses and reduce objections from family members. As there is a paucity of Hong Kong research data on advance directive, more local studies on their implementation in different patient groups and at different end-stage illnesses are necessary. Apart from studying the uptake and associated contributory factors for advance directives, they should investigate the effectiveness of advance directive implementation strategies and their impact on hospitalisations and medical expenditure. Overseas studies showed that persons who had assigned a durable power of attorney for health care were less likely to die in a hospital or receive all care possible. Moreover, the use of advance directives resulted in a reduction in health care expenditures, when the directives specified the limits to end-of-life care. However, this cannot be taken for granted. Hong Kong has a very different public health care financing system to the US, for which reason they may not be the same. Nonetheless, when future studies are designed on advance directives in Hong Kong, assessment of these outcomes should also be incorporated.

2. Not to Be Accepted

a. Not an Option in Psychiatric Care

While there are a number of reasons to support advance directives, there are a number of reasons given for why they are not followed. Physician and bioethicist Abraham Rudnick is a professor in the Department of Behavioral Sciences at Tel Aviv

University School of Medicine in Israel. In what follows, Rudnick addresses the issue of treatment refusal in psychiatric care. In countries such as Israel and the United States, a psychiatric patient has a right to refuse treatment. Yet a psychiatric condition can undermine a patient's ability to make decisions, including consenting to or refusing clinical treatment. It can undermine a patient's ability to choose because a psychiatric condition may compromise the ability to receive and understand information, to consent to a plan, and to act on such information and the chosen plan. The question arises, then, can a psychiatric patient understand the implications of refusing treatment that may lead to death? In what follows, Rudnick calls for greater discussion of treatment refusal in the case of those who are severely depressed. In particular, he would like to see a rethinking of the criteria developed for what constitutes "coherence of personal preferences" (2002, 151) in cases of treatment refusal in psychiatry.

Reading: "Depression and the Competence to Refuse Psychiatric Treatment" (Rudnick 2002, 151, 155). Reproduced with the permission of BMJ Publishing Group Ltd.

[p. 151]

Abstract

Individuals with major depression may benefit from psychiatric treatment, yet they may refuse such treatment, sometimes because of their depression. Hence the question is raised whether such individuals are competent to refuse psychiatric treatment. The standard notion of competence to consent to treatment, which refers to expression of choice, understanding of medical information, appreciation of the personal relevance of this information, and logical reasoning, may be insufficient to address this question. This is so because major depression may not impair these four abilities while it may disrupt coherence of personal preferences by changing them. Such change may be evaluated by comparing the treatment preferences of the individual during the depression to his or her treatment preferences during normal periods. If these preferences are consistent, they should be respected. If they are not consistent, or past treatment preferences that were arrived at competently cannot be established, treatment refusal may have to be overridden or ignored so as to alleviate the depression and then determine the competent treatment decision of the individual. Further study of the relation between depression and competence to refuse or consent to psychiatric treatment is required. . . .

[p. 155]

Conclusion

The question of the competence of depressed patients to consent to or refuse treatment for their depression has not been sufficiently addressed to date. It raises difficulties for the standard notion of competence to consent to treatment, as the latter does not address preferences that serve as premises of decision making, whereas depression—particularly major depression without psychotic features—may considerably impact on decision making through its impact on such preferences. Thus, major depression may sometimes change current treatment preferences of patients so that they are not consistent with their past treatment preferences, and then a determination has to be made as to which preferences of the patient to follow (so long as it can be shown that the inconsistency is not due to other changes, such as changes in circumstances which may provide reasonable grounds for change of mind). In such cases, when past treatment preferences of the patient from periods without depression can be established, it is suggested that they should override current treatment preferences of the patient. If past treatment preferences cannot be established with sufficient certitude, a therapeutic trial of treatment for the depression may be justified, even if the patient currently refuses the treatment, as in such circumstances the determination of competence during the depression may be possible only after successful treatment of the depression, when it can be evaluated whether the treatment preferences during and after the depression are consistent.

Further study of this suggested addition to the notion of competence is required. Empirical research of changes in treatment preferences of severely depressed (but not psychotic) patients, following successful treatment of their depression, may shed light on the question whether severe depression without psychotic features usually disrupts competence to refuse or consent to psychiatric treatment. Legal research could explore the possibility of formalising such a suggestion, in particular the potentially controversial idea of a therapeutic trial that may override/ignore current patient treatment preferences in order to assess competence. And social research could clarify the attitudes of the various parties involved, such as depressed patients, their families, clinicians, policy makers, and the general public, towards such a suggestion, so as to determine some of the practical obstacles such a suggestion may confront. Finally, conceptual investigation of the application or modification of this suggestion so as to address more chronic forms of depression, such as dysthymia, may test its limits in that depression-related preferences in such forms of depression may be so ingrained and longstanding that successful treatment may be viewed as disrupting coherence of preferences rather than reinstating it, which might make such patients on this account competent when depressed and incompetent if and when they are not depressed! Be that as it may, some such suggestion may be helpful in evaluating the competence of depressed patients to refuse or consent to psychiatric treatment.

b. Not an Option for Parents to Decide for Children

In order to sign a living will or durable power of attorney for health care, one must be an adult, or eighteen years of age or older. This is the case because adults are seen to be capable of making an informed decision to refuse treatment. They are seen to understand what is involved in refusing treatment. Those eighteen years of age and older who are parents are also legally able to make decisions for their children. Here is where some issues concerning treatment refusal are raised. There are numerous discussions online and in the literature about whether a parent should be able to refuse treatment for a child. Carol Forsloff, a contributor to *Yahoo! Voice*, addresses this issue: "Children are vulnerable, and the law supports the view that the child, when it comes to health, is part of an overall concern that the state has for the protection and welfare of those who are unable to protect or defend themselves" (2009, ¶3). In what follows, Forsloff addresses the issue of parents refusing treatment for children. She focuses on cases of treatment refusal based on religious beliefs.

Reading: "Religion and the Refusal of Medical Treatment: Rights and Responsibilities—Adults Can Say No for Themselves but Not for Children" (Forsloff 2009)

There are certain religious groups that don't believe in traditional medicine and instead turn to prayer and other means to heal themselves. Some of these people refuse medical treatment for their children, and some of these children die as a consequence. Given these issues, it is important to explore what a person can or cannot do to avoid having trouble with the law and what some of the consequences might be for someone refusing to give medical care to a child.

The law upholds an adult's right to refuse medical care. That's because of the freedom of religion clause in the Constitution. On the other hand, that person must be adjudged mentally competent to be able to make a reasonable decision. So people who are members of Christian Science or Jehovah's Witnesses, or fundamental Christians may refuse medical treatment so long as they are capable of making a rational decision about it.

The same does not hold true in the case of children. Children are vulnerable, and the law supports the view that the child, when it comes to health, is part of an overall concern that the state has for the protection and welfare of those who are unable to protect or defend themselves. Therefore, if a parent refuses medical care and the child dies, the parent may face legal consequences as a result.

The *New York Times* has detailed some of the cases of parents who have refused to get their children proper medical care and the children died. One of them, Kara Neuman, who was diabetic, died as a result of not having care for her diabetes. The parents have been found culpable.

One author has written specifically on the subject and has provided valuable information that can be reviewed that looks at certain cases and the ambiguities of some of them. The issues that become controversial are those that involve whether or not the therapy suggested by a doctor or medical care facility has a likely chance of saving a life and what the chances are of that occurring vs. doing potential harm.

This is the dilemma as stated by one group: "When parents wish to withhold or discontinue standard treatment in a child with cancer that has a likelihood of long-term cure, referral to the local child protection agency is indicated because a parent's inability to provide adequate care for a child is a criminal offense. The difficulty lies in defining a threshold of therapeutic success where treatment should be initiated against the wishes of the parent or guardian. This problem has been challenging for the legal system to resolve. Courts will order treatment over parental objections for conditions that are immediately life threatening, such as antibiotics for bacterial meningitis or insulin injections for diabetes mellitus; however, when the disease does not cause imminent harm but there are significant risks involved with the treatment, the decision of the court is less predictable."

So the answer to the issues regarding treatment of a child is less clear than for an adult, but the concerns about known risks that can lead to death are defined sharply as a criminal offense when medical treatment is withheld.

How all of this will affect vaccines and autism will be a cause of likely debate, given the evidence that has been found that vaccines don't cause autism and the refusal of parents to provide vaccinations that, in some cases, have led to the death of the children concerned, is something the courts will have to deal and provide precedent.

The answer is likely that same admonition to parents that doctors have: do no harm.

Note 1: Here, "traditional medicine" refers to scientific medicine that relies on biomedical theory and technology to diagnose and treat patients.

c. Not an Option for Some Ethnic Groups

More work is needed on end-of-life decision making in order to understand how individuals, families, and communities make the decisions they do. Professor Jung Kwak teaches social work at the University of Wisconsin at Milwaukee, and Professor William Haley teaches in the School of Aging Studies at the University of South Florida at Tampa. Both scholars have been researching preferences for end-of-life care among ethnically diverse groups for many years. In what follows, Kwak and Haley observe, "Non-White racial or ethnic groups generally lacked knowledge of advance directives and were less likely than Whites to support advance directives. African Americans were consistently found to prefer the use of life support; Asians and Hispanics were more likely to prefer family-centered decision making than other racial or ethnic groups. Variations within groups existed and were related to cultural values, demographic characteristics, level of acculturation, and knowledge of end-of-life treatment options" (2005, 634). Kwak and Haley draw our attention to the need to explore cultural influences on end-of-life decision making. Not all groups of people assume that the individual is a primary decision maker who makes choices based on rational, sequential assessment. Many hold that important decisions need to be made by, for example, the head of a clan, the family as a collective unit, or some other group. What follows is the abstract for their essay in the *Gerontologist*. It leads us to question our assumptions about who decides and how at the end of life, and in doing so it opens up avenues of discussion for considering ways to help others with end-of-life decisions.

Reading: "Current Research Findings on End-of-Life Decision Making among Racially or Ethnically Diverse Groups" (Kwak and Haley 2005, 634). Reproduced with the permission of Oxford University Press.

Abstract

Purpose
We reviewed the research literature on racial or ethnic diversity and end-of-life decision making in order to identify key findings and provide recommendations for future research.

Design and Methods
We identified 33 empirical studies in which race or ethnicity was investigated as either a variable predicting treatment preferences or choices, where racial or ethnic groups were compared in their end-of-life decisions, or where the end-of-

life decision making of a single minority group was studied in depth. We conducted a narrative review and identified four topical domains of study: advance directives; life support; disclosure and communication of diagnosis, prognosis, and preferences; and designation of primary decision makers.

Results
Non-White racial or ethnic groups generally lacked knowledge of advance directives and were less likely than Whites to support advance directives. African Americans were consistently found to prefer the use of life support; Asians and Hispanics were more likely to prefer family-centered decision making than other racial or ethnic groups. Variations within groups existed and were related to cultural values, demographic characteristics, level of acculturation, and knowledge of end-of-life treatment options. Common methodological limitations of these studies were lack of theoretical framework, use of cross-sectional designs, convenience samples, and self-developed measurement scales.

Implications
Although the studies are limited by methodological concerns, identified differences in end-of-life decision-making preference and practice suggest that clinical care and policy should recognize the variety of values and preferences found among diverse racial or ethnic groups. Future research priorities are described to better inform clinicians and policy makers about ways to allow for more culturally sensitive approaches to end-of-life care.

C. REFLECTIONS: LIVING WILL AND MEDICAL DURABLE POWER OF ATTORNEY CONSENT FORMS

1.(*) In your view, does a patient have a right to refuse clinical treatment at the end of life? In your response, draw from philosophical, clinical, and/or sacred text interpretations and give details.

2.(*) In your view, does a parent have a right to refuse treatment for a child at the end of the child's life? In your response, draw from philosophical, clinical, and/or sacred text interpretations and give details.

3.   This is a continuation of your advance planning in preparation for death:
    a. Find the "living will" in the state in which you reside. You may find one online (e.g., google "Colorado Living Will"). You may also find one at the information desk of a local hospital. Do not pay for one; you should be able

to find one for free. Complete the form. Please note: This is a legal document in the state in which you reside. Should you not want this to be a legal document, write or type "void" across after you complete it.

b. Find the "medical durable power of attorney" in the state in which you reside. You may find one online (e.g., google "Colorado Medical Durable Power of Attorney"). You may also find one at the information desk of a local hospital. Do not pay for one; you should be able to find one for free. Complete the form. Please note: This is a legal document in the state in which you reside. Should you not want this to be a legal document, write or type "void" across it after you complete it.

D. FURTHER READINGS

Berlinger, Nancy, Bruce Jennings, and Susan M. Wolf. 2013. *The Hastings Center Guidelines for Decisions on Life-Sustaining Treatment and Care near the End of Life*. 2nd ed. New York: Oxford University Press.

Byrock, Ira. 1998. *Dying Well*. New York: Riverhead Books.

———. 2012. *The Best Care Possible*. New York: Avery.

Callanan, Maggie, and Patricia Kelley. [1992] 2012. *Final Gifts: Understanding the Special Awareness, Needs, and Communications of the Dying*. New York: Simon and Schuster.

Chu, Leung Wing, James K. H. Luk, Elsie Hui, Patrick K. C. Chiu, Cherry S. Y. Chan, Fiona Kwan, Timothy Kwok, Diana Lee, and Jean Woo. 2011. "Advance Directive and End-of-Life Care Preferences among Chinese Nursing Home Residents in Hong Kong." *Journal of the American Medical Directors Association* 12 (2): 143–52.

Colby, W. H. 2006. *Unplugged: Reclaiming Our Right to Die in America*. New York: Amacom.

"Euthanasia and Physician-Assisted Suicide." 2016. *Religious Tolerance*. www.religioustolerance.org/euthanas.htm#.

Kessler, David. 2007. *The Needs of the Dying: A Guide for Bringing Hope, Comfort, and Love to Life's Final Chapter*. New York: Harper Perennial.

Robert Wood Foundation. 2014. *"Five Wishes": An Easy Advance Directive Promotes Dialogue on End-of-Life Care*. http://pweb1.rwjf.org/reports/grr/029110s.htm.

Chapter 14

To Be Hastened or Not

The Case of Physician-Assisted Suicide

A. CONTEXT

Today, much attention is given to the topic of whether it is permissible to hasten death through the means of physician-assisted suicide. The discussion emerges because, as seen in the last chapter, contemporary medicine has the ability to treat disease in many effective ways, thereby prolonging the life of those who have access to such medical interventions. When prolongation of life is not in concert with a patient's choice or welfare and a patient requests assistance in dying, a patient might seek assistance in death through physician-assisted suicide in a state or country in which it is legally permissible.

Increasingly, states and countries are entertaining legislation and policies that reflect their support or disapproval of physician-assisted suicide. As of this writing, in the United States, physician-assisted suicide is legal in the states of Oregon (1994), Washington (2008), Vermont (2013), Colorado (2016), District of Columbia (2016), and Hawaii (2018). Montana (2009) does not have a law against physician-assisted suicide, but the court ruling in *Baxter vs. Montana* protects physicians from prosecution if they write a prescription for lethal medication on the request of a terminally ill patient. California passed physician-assisted suicide legislation in 2015 but reversed it in 2018. Physician-assisted suicide is legal in the countries of Netherlands (2002), Belgium (2002), Switzerland (2002), Luxembourg (2008), Canada (2015), and Colombia (2015). Belgium allows physician-assisted suicide for consenting minors

(McDonald-Gibson 2014). Sweden and Norway do not prosecute physician-assisted suicide cases. Physician-assisted suicide is generally illegal in all countries in Asia and the Pacific ("Euthanasia and Physician-Assisted Suicide" 2018). Given that state and national laws concerning physician-assisted suicide are changing each year, readers are encouraged to update this information after major elections.

What follows is a discussion of death from the perspective of whether it is permissible to hasten it through the means of physician-assisted suicide. As one can imagine, there are arguments on both sides, as the *Ethics in Clerkships* article below indicates. On the one hand, the current legislation in the Netherlands permits physician-assisted suicide when a competent adult patient chooses it and a series of conditions are met. The Oregon Death with Dignity Act allows a competent adult patient with a terminal condition to choose physician assistance to end his or her life as long as certain conditions are met. On the other hand, the current Criminal Code in China (Articles 232 and 233) and the teachings of the National Association of Evangelicals prohibit assisted suicide. The selections that follow illustrate a spectrum of views on the ethics of physician-assisted suicide and particular understandings of death that are valued and disvalued.

B. PERSPECTIVES

1. To Be Hastened

a. To Be Assisted

Arguments in favor of physician-assisted suicide abound today. In what follows, a short summary posted on the *Ethics in Clerkships* website summarizes reasons in favor of physician-assisted suicide. One finds a mix of justifications that appeal to patient welfare, patient autonomy, and justice. Appeal to patient welfare is found in items a, d, and e below. In particular, protecting patients from "lingering, painful deaths" ("Topics" 2013) is a goal with physician-assisted suicide. Appeal to patient autonomy is found in items b, c, and f below. In particular, allowing patients to make decisions about their own life and death is a priority for advocates of physician-assisted suicide. Appeal to justice is found in item f. In particular, the state's role in protecting a patient's refusal to prolong life is mentioned.

Reading: "Topics: Physician-Assisted Suicide," on *Ethics in Clerkships* website ("Topics" 2013)

As with the case against physician-assisted suicide, certain arguments are repeated in these movements favoring physician-assisted suicide:

a. it protects people who do not want to suffer lingering, painful deaths
b. it is in keeping with respect for patient autonomy
c. it is defensible as policy because it respects social diversity
d. it protects against physician paternalism and unwanted treatment
e. it protects against debilitating conditions not easily managed by medicine
f. the state has no interest in forcing the prolongation of life of someone in pain who wants to die

There is no moral or legal support for physician assistance in any kind of involuntary death. No serious advocate of physician-assisted suicide has argued that physicians must take part in assisting in death. Proponents of physician-assisted suicide recognize the right of individual physicians to decline to participate for religious or moral reasons. It is though sometimes argued that physicians should, regardless of their own moral views about assisted death, offer referral of patients to physicians who will help them in the desired way.

b. Voluntary and Permissibly Assisted

In 1985, the Netherlands State Commission on Euthanasia adopted initial legislation supporting the termination of life on request. In 2002, review procedures for termination on request and assisted suicide were developed. Euthanasia in the Netherlands has officially been defined as "intentionally terminating another person's life at that person's request." As required, the request must be "voluntary and well-considered" ("Termination" 2002, II.2.1.a), the patient's suffering must be "lasting and unbearable" ("Termination" 2002, II.2.1.b), the patient must be informed of his or her clinical condition and its prognosis ("Termination" 2002, II.2.1.c), the patient must hold "the conviction that there [is] no other reasonable solution for the situation" ("Termination" 2002, II.2.1.d) he or she is in, a second opinion must be sought ("Termination" 2002, II.2.1.e), the request must be made from an adult 18 years or older (with exceptions made for minors 12 to 16 years of age) ("Termination" 2002, II.2.2), and the request must be reviewed by a "regional review committee" ("Termination" 2002, III). What follows is chapter I through part of chapter III of the 2002 legislation for permissible termination of life on request and assisted suicide adopted by the Dutch government.

Reading: "Termination of Life on Request and Assisted Suicide (Review Procedures) Act" [Netherlands] (2002)

This Act entered into force on April 1, 2002.

Review procedures of termination of life on request and assisted suicide and amendment to the Penal Code (Wetboek van Strafrecht) and the Burial and Cremation Act (Wet op de lijkbezorging)

We Beatrix, by the grace of God, Queen of the Netherlands, Princess of Oranje-Nassau, etc., etc. etc.

Greetings to all who shall see or hear these presents! Be it known:

Whereas We have considered that it is desired to include a ground for exemption from criminal liability for the physician who with due observance of the requirements of due care to be laid down by law terminates a life on request or assists in a suicide of another person, and to provide a statutory notification and review procedure;

We, therefore, having heard the Council of State, and in consultation with the States General, have approved and decreed as We hereby approve and decree:

Chapter I. Definitions of Terms

Article 1

For the purposes of this Act:

a. Our Ministers mean the Ministers of Justice and of Health, Welfare and Sports;

b. assisted suicide means intentionally assisting in a suicide of another person or procuring for that other person the means referred to in Article 294 second paragraph, second sentence of the Penal code;

c. the physician means the physician who according to the notification has terminated a life on request or assisted in a suicide;

d. the consultant means the physician who has been consulted with respect to the intention by the physician to terminate a life on request or to assist in a suicide;

e. the providers of care mean the providers of care referred to in Article 446 first paragraph of Book 7 of the Civil Code (Burgerlijk Wetboek);

f. the committee means a regional review committee referred to in Article 3;

g. the regional inspector means the regional inspector of the Health Care Inspectorate of the Public Health Supervisory Service.

Chapter II. Requirements of Due Care

Article 2

1. The requirements of due care, referred to in Article 293 second paragraph
    Penal Code mean that the physician:

    a. holds the conviction that the request by the patient was voluntary and
        well-considered,

    b. holds the conviction that the patient's suffering was lasting and unbear-
        able,

    c. has informed the patient about the situation he was in and about his
        prospects,

    d. and the patient hold the conviction that there was no other reasonable
        solution for the situation he was in,

    e. has consulted at least one other, independent physician who has seen
        the patient and has given his written opinion on the requirements of due
        care, referred to in parts a–d, and

    f. has terminated a life or assisted in a suicide with due care.

2. If the patient aged sixteen years or older is no longer capable of expressing
    his will, but prior to reaching this condition was deemed to have a reason-
    able understanding of his interests and has made a written statement con-
    taining a request for termination of life, the physician may grant this request.
    The requirements of due care, referred to in the first paragraph, apply mu-
    tatis mutandis.

3. If the minor patient has attained an age between sixteen and eighteen years
    and may be deemed to have a reasonable understanding of his interests,
    the physician may grant the patient's request for termination of life or as-
    sisted suicide, after the parent or the parents exercising parental authority
    and/or his guardian have been involved in the decision process.

4. If the minor patient is aged between twelve and sixteen years and may be
    deemed to have a reasonable understanding of his interests, the physician
    may grant the patient's request, provided always that the parent or the par-
    ents exercising parental authority and/or his guardian agree with the termina-
    tion of life or the assisted suicide. The second paragraph applies mutatis mu-
    tandis.

Chapter III. The Regional Review Committees for Termination of Life on Request
and Assisted Suicide.

Paragraph 1: Establishment, composition and appointment

Article 3

1. There are regional committees for the review of notifications of cases of
    termination of life on request and assistance in a suicide as referred to in

Article 293 second paragraph or 294 second paragraph second sentence, respectively, of the Penal Code.

2. A committee is composed of an uneven number of members, including at any rate one legal specialist, also chairman, one physician and one expert on ethical or philosophical issues [*Ed. note: in original, the word translated as "philosophical issues" is* zingevingsvraagstukken *(meaning questions, or questions regarding the prerequisites for a meaningful life)*]. The committee also contains deputy members of each of the categories listed in the first sentence.

Article 4

1. The chairman and the members, as well as the deputy members are appointed by Our Ministers for a period of six years. They may be re-appointed one time for another period of six years.

2. A committee has a secretary and one or more deputy secretaries, all legal specialists, appointed by Our Ministers. The secretary has an advisory role in the committee meetings.

3. The secretary may solely be held accountable by the committee for his activities for the committee.

Paragraph 2: Dismissal

Article 5

Our Ministers may at any time dismiss the chairman and the members, as well as the deputy members at their own request.

Article 6

Our Ministers may dismiss the chairman and the members, as well as the deputy members for reasons of unsuitability or incompetence or for other important reasons.

Paragraph 3: Remuneration

Article 7

The chairman and the members, as well as the deputy members receive a holiday allowance as well as a reimbursement of the travel and accommodation expenses according to the existing government scheme insofar as these expenses are not otherwise reimbursed from the State Funds.

Paragraph 4: Duties and powers

Article 8

1. The committee assesses on the basis of the report referred to in Article 7 second paragraph of the Burial and Cremation Act whether the physician who has terminated a life on request or assisted in a suicide has acted in accordance with the requirements of due care, referred to in Article 2.

2. The committee may request the physician to supplement his report in writing or verbally, where this is necessary for a proper assessment of the physician's actions.

3. The committee may make enquiries at the municipal autopsist, the consultant or the providers of care involved where this is necessary for a proper assessment of the physician's actions.

Article 9

1. The committee informs the physician within six weeks of the receipt of the report referred to in Article 8 first paragraph in writing of its motivated opinion.

2. The committee informs the Board of Procurators General and the regional health care inspector of its opinion:

    a. if the committee is of the opinion that the physician has failed to act in accordance with the requirements of due care, referred to in Article 2; or

    b. if a situation occurs as referred to in Article 12, final sentence of the Burial and Cremation Act.
    The committee shall inform the physician of this.

3. The term referred to in the first paragraph may be extended one time by a maximum period of six weeks. The committee shall inform the physician of this.

4. The committee may provide a further, verbal explanation on its opinion to the physician. This verbal explanation may take place at the request of the committee or at the request of the physician.

Article 10

The committee is obliged to provide all information to the public prosecutor, at his request, which he may need:

1. for the benefit of the assessment of the physician's actions in the case referred to in Article 9 second paragraph; or

2. for the benefit of a criminal investigation.

The committee shall inform the physician of any provision of information to the public prosecutor.

c. Requested

On October 27, 1997, Oregon enacted the Death with Dignity Act, which allows terminally ill Oregonians to end their lives through the voluntary self-administration of lethal medications, expressly prescribed by a physician for that purpose. As the legislation states, "An adult who is capable, is a resident of Oregon, and has been determined by the attending physician and consulting physician to be suffering from a terminal disease, and who has voluntarily expressed his or her wish to die, may make a written request for medication for the purpose of ending his or her life in a humane and dignified manner in accordance with ORS 127.800 to 127.897" ("Oregon" 2013, 127.805 s.2.01). The patient must be an Oregonian resident and suffer from a terminal illness. The Oregon Death with Dignity Act requires the Oregon Health Authority to collect information about patients and physicians who participate in the act and to publish an annual statistical report. What follows is a large part of the legislation governing physician-assisted suicide in Oregon.

Reading: "Oregon Death with Dignity Act" (2013, 127.800–127.860). Reproduced with the permission of the Oregon Health Authority.

127.800 s.1.01. Definitions.
The following words and phrases, whenever used in ORS 127.800 to 127.897, have the following meanings:
(1) "Adult" means an individual who is 18 years of age or older.
(2) "Attending physician" means the physician who has primary responsibility for the care of the patient and treatment of the patient's terminal disease.
(3) "Capable" means that in the opinion of a court or in the opinion of the patient's attending physician or consulting physician, psychiatrist or psychologist, a patient has the ability to make and communicate health care decisions to health care providers, including communication through persons familiar with the patient's manner of communicating if those persons are available.
(4) "Consulting physician" means a physician who is qualified by specialty or experience to make a professional diagnosis and prognosis regarding the patient's disease.
(5) "Counseling" means one or more consultations as necessary between a state licensed psychiatrist or psychologist and a patient for the purpose of determining that the patient is capable and not suffering from a psychiatric or psychological disorder or depression causing impaired judgment.
(6) "Health care provider" means a person licensed, certified or otherwise authorized or permitted by the law of this state to administer health care or dispense medication in the ordinary course of business or practice of a profession, and includes a health care facility.
(7) "Informed decision" means a decision by a qualified patient, to request and obtain a prescription to end his or her life in a humane and dignified manner, that

is based on an appreciation of the relevant facts and after being fully informed by the attending physician of:

   (a) His or her medical diagnosis;

   (b) His or her prognosis;

   (c) The potential risks associated with taking the medication to be prescribed;

   (d) The probable result of taking the medication to be prescribed; and

   (e) The feasible alternatives, including, but not limited to, comfort care, hospice care and pain control.

(8) "Medically confirmed" means the medical opinion of the attending physician has been confirmed by a consulting physician who has examined the patient and the patient's relevant medical records.

(9) "Patient" means a person who is under the care of a physician.

(10) "Physician" means a doctor of medicine or osteopathy licensed to practice medicine by the Board of Medical Examiners for the State of Oregon.

(11) "Qualified patient" means a capable adult who is a resident of Oregon and has satisfied the requirements of ORS 127.800 to 127.897 in order to obtain a prescription for medication to end his or her life in a humane and dignified manner.

(12) "Terminal disease" means an incurable and irreversible disease that has been medically confirmed and will, within reasonable medical judgment, produce death within six months. [1995 c.3 s.1.01; 1999 c.423 s.1]

(Written Request for Medication to End One's Life in a Humane and Dignified Manner)

(Section 2)

127.805 s.2.01. Who may initiate a written request for medication.

(1) An adult who is capable, is a resident of Oregon, and has been determined by the attending physician and consulting physician to be suffering from a terminal disease, and who has voluntarily expressed his or her wish to die, may make a written request for medication for the purpose of ending his or her life in a humane and dignified manner in accordance with ORS 127.800 to 127.897.

(2) No person shall qualify under the provisions of ORS 127.800 to 127.897 solely because of age or disability. [1995 c.3 s.2.01; 1999 c.423 s.2]

127.810 s.2.02. Form of the written request.

(1) A valid request for medication under ORS 127.800 to 127.897 shall be in substantially the form described in ORS 127.897, signed and dated by the patient and witnessed by at least two individuals who, in the presence of the patient,

attest that to the best of their knowledge and belief the patient is capable, acting voluntarily, and is not being coerced to sign the request.

(2) One of the witnesses shall be a person who is not:

(a) A relative of the patient by blood, marriage or adoption;

(b) A person who at the time the request is signed would be entitled to any portion of the estate of the qualified patient upon death under any will or by operation of law; or

(c) An owner, operator or employee of a health care facility where the qualified patient is receiving medical treatment or is a resident.

(3) The patient's attending physician at the time the request is signed shall not be a witness.

(4) If the patient is a patient in a long term care facility at the time the written request is made, one of the witnesses shall be an individual designated by the facility and having the qualifications specified by the Oregon Health Authority by rule. [1995 c.3 s.2.02]

(Safeguards)

(Section 3)

127.815 s.3.01. Attending physician responsibilities.

(1) The attending physician shall:

(a) Make the initial determination of whether a patient has a terminal disease, is capable, and has made the request voluntarily;

(b) Request that the patient demonstrate Oregon residency pursuant to ORS 127.860;

(c) To ensure that the patient is making an informed decision, inform the patient of:

(A) His or her medical diagnosis;

(B) His or her prognosis;

(C) The potential risks associated with taking the medication to be prescribed;

(D) The probable result of taking the medication to be prescribed; and

(E) The feasible alternatives, including, but not limited to, comfort care, hospice care and pain control;

(d) Refer the patient to a consulting physician for medical confirmation of the diagnosis, and for a determination that the patient is capable and acting voluntarily;

(e) Refer the patient for counseling if appropriate pursuant to ORS 127.825;

(f) Recommend that the patient notify next of kin;

(g) Counsel the patient about the importance of having another person present when the patient takes the medication prescribed pursuant to ORS 127.800 to 127.897 and of not taking the medication in a public place;

(h) Inform the patient that he or she has an opportunity to rescind the request at any time and in any manner, and offer the patient an opportunity to rescind at the end of the 15 day waiting period pursuant to ORS 127.840;

(i) Verify, immediately prior to writing the prescription for medication under ORS 127.800 to 127.897, that the patient is making an informed decision;

(j) Fulfill the medical record documentation requirements of ORS 127.855;

(k) Ensure that all appropriate steps are carried out in accordance with ORS 127.800 to 127.897 prior to writing a prescription for medication to enable a qualified patient to end his or her life in a humane and dignified manner; and

(L)(A) Dispense medications directly, including ancillary medications intended to facilitate the desired effect to minimize the patient's discomfort, provided the attending physician is registered as a dispensing physician with the Board of Medical Examiners, has a current Drug Enforcement Administration certificate and complies with any applicable administrative rule; or

(B) With the patient's written consent:

(i) Contact a pharmacist and inform the pharmacist of the prescription; and

(ii) Deliver the written prescription personally or by mail to the pharmacist, who will dispense the medications to either the patient, the attending physician or an expressly identified agent of the patient.

(2) Notwithstanding any other provision of law, the attending physician may sign the patient's death certificate. [1995 c.3 s.3.01; 1999 c.423 s.3]

127.820 s.3.02. Consulting physician confirmation.

Before a patient is qualified under ORS 127.800 to 127.897, a consulting physician shall examine the patient and his or her relevant medical records and confirm, in writing, the attending physician's diagnosis that the patient is suffering from a terminal disease, and verify that the patient is capable, is acting voluntarily and has made an informed decision. [1995 c.3 s.3.02]

127.825 s.3.03. Counseling referral.

If in the opinion of the attending physician or the consulting physician a patient may be suffering from a psychiatric or psychological disorder or depression causing impaired judgment, either physician shall refer the patient for counseling. No medication to end a patient's life in a humane and dignified manner shall be prescribed until the person performing the counseling determines that the patient is not suffering from a psychiatric or psychological disorder or depression causing impaired judgment. [1995 c.3 s.3.03; 1999 c.423 s.4]

127.830 s.3.04. Informed decision.
No person shall receive a prescription for medication to end his or her life in a humane and dignified manner unless he or she has made an informed decision as defined in ORS 127.800 (7). Immediately prior to writing a prescription for medication under ORS 127.800 to 127.897, the attending physician shall verify that the patient is making an informed decision. [1995 c.3 s.3.04]

127.835 s.3.05. Family notification.
The attending physician shall recommend that the patient notify the next of kin of his or her request for medication pursuant to ORS 127.800 to 127.897. A patient who declines or is unable to notify next of kin shall not have his or her request denied for that reason. [1995 c.3 s.3.05; 1999 c.423 s.6]

127.840 s.3.06. Written and oral requests.
In order to receive a prescription for medication to end his or her life in a humane and dignified manner, a qualified patient shall have made an oral request and a written request, and reiterate the oral request to his or her attending physician no less than fifteen (15) days after making the initial oral request. At the time the qualified patient makes his or her second oral request, the attending physician shall offer the patient an opportunity to rescind the request. [1995 c.3 s.3.06]

127.845 s.3.07. Right to rescind request.
A patient may rescind his or her request at any time and in any manner without regard to his or her mental state. No prescription for medication under ORS 127.800 to 127.897 may be written without the attending physician offering the qualified patient an opportunity to rescind the request. [1995 c.3 s.3.07]

127.850 s.3.08. Waiting periods.
No less than fifteen (15) days shall elapse between the patient's initial oral request and the writing of a prescription under ORS 127.800 to 127.897. No less than 48 hours shall elapse between the patient's written request and the writing of a prescription under ORS 127.800 to 127.897. [1995 c.3 s.3.08]

127.855 s.3.09. Medical record documentation requirements.
The following shall be documented or filed in the patient's medical record:
(1) All oral requests by a patient for medication to end his or her life in a humane and dignified manner;

(2) All written requests by a patient for medication to end his or her life in a humane and dignified manner;

(3) The attending physician's diagnosis and prognosis, determination that the patient is capable, acting voluntarily and has made an informed decision;

(4) The consulting physician's diagnosis and prognosis, and verification that the patient is capable, acting voluntarily and has made an informed decision;

(5) A report of the outcome and determinations made during counseling, if performed;

(6) The attending physician's offer to the patient to rescind his or her request at the time of the patient's second oral request pursuant to ORS 127.840; and

(7) A note by the attending physician indicating that all requirements under ORS 127.800 to 127.897 have been met and indicating the steps taken to carry out the request, including a notation of the medication prescribed. [1995 c.3 s.3.09]

127.860 s.3.10. Residency requirement.

Only requests made by Oregon residents under ORS 127.800 to 127.897 shall be granted. Factors demonstrating Oregon residency include but are not limited to:

(1) Possession of an Oregon driver license;

(2) Registration to vote in Oregon;

(3) Evidence that the person owns or leases property in Oregon; or

(4) Filing of an Oregon tax return for the most recent tax year. [1995 c.3 s.3.10; 1999 c.423 s.8]

2. Not to Be Hastened

a. Not to Be Assisted

Arguments against physician-assisted suicide are not difficult to find. In what follows, a short summary posted on the *Ethics in Clerkships* website summarizes reasons against physician-assisted suicide. One finds a mix of justifications that appeal to patient welfare, patient autonomy, and justice. Appeal to patient welfare is found in items b, c, and f below. In particular, physician-assisted suicide is seen to be "incompatible with the healing goals of medicine" ("Topics" 2012). Appeal to patient autonomy is found in point d below. In particular, a question is raised about whether the choice to request physician-assisted suicide is autonomous. Appeal to justice is found in item f below. In particular, concerns about the indiscriminate killing of the vulnerable are raised. Not represented here are a host of religious and spiritual views from Western and Asian worldviews regarding the impermissibility of assisting the death of another. Readers might consider these as well.

Reading: "Topics: Physician-Assisted Suicide," on *Ethics in Clerkships* website ("Topics" 2013)

There are a number of arguments that are repeated in the argument against physician-assisted suicide:

a.   suicide is wrong in and of itself even for the ill
b.   it is incompatible with the healing goals of medicine
c.   given appropriate palliative care, it is unnecessary
d.   requests for death are induced by poor care and/or unrecognized psycho-
     logical needs
e.   the practice damages physicians by desensitizing them to human needs
f.   it leads down a slippery slope to indiscriminate killing of the ill, weak, and
     disabled, among others

Taken either separately or in some combination, these arguments are often found powerful and convincing by physicians, moralists, and the public alike.

At present, the A.M.A. declares the profession entirely opposed to physician-assisted suicide: "Physician-assisted suicide is fundamentally inconsistent with the physician's professional role." Instead of physician involvement in assisting the death of patients, the A.M.A. counsels physicians to tend assiduously to the pain and discomfort of the dying. "The use of more aggressive comfort care measures, including greater reliance on hospice care, can alleviate the physical and emotional suffering that dying patients experience." [A.M.A. Code of Medical Ethics Reports 1994 (V), Report 59.] Some psychiatrists even hold that requests by patients for assistance in death are prima facie evidence of incompetence; these requests should not therefore be honored under any circumstances.

b. Illegal

Currently, the Criminal Law of the People's Republic of China (Articles 232 and 233) prohibits assisted suicide, including physician-assisted suicide. Journalist Wendy Zeldin sets the stage for an interpretation of Articles 232 and 233 of the Chinese Criminal Law with a story about an assisted suicide case that gained attention in

China in 2010. As the story goes, Zhong Yichun assisted the death of his friend Zeng Qianxiang, who wished to commit suicide. According to the Criminal Law of the People's Republic of China, assisting the death of another is considered a violation of law and punishable by imprisonment. As Zeldin reports, "The case has given rise to renewed nation-wide debate on euthanasia, which is banned under current Chinese law; there is no provision permitting assisted suicide" (¶3). What follows are Articles 232 and 233 of the Criminal Law of the People's Republic of China and Zeldin's recounting of a story of assisted suicide in China. Readers may note that, although the relevant Criminal Code provision has not been amended as of this printing, laws and regulations can change.

Reading: "Criminal Law of the People's Republic of China" (Criminal Law of the People's Republic of China 2013, Articles 232 and 233)

Chapter IV Crimes of Infringing upon Citizens' Right of the Person and Democratic Rights

Article 232 Whoever intentionally commits homicide shall be sentenced to death, life imprisonment or fixed-term imprisonment of not less than 10 years; if the circumstances are relatively minor, he shall be sentenced to fixed-term imprisonment of not less than three years but not more than 10 years.

Article 233 Whoever negligently causes death to another person shall be sentenced to fixed-term imprisonment of not less than three years but not more than seven years; if the circumstances are relatively minor, he shall be sentenced to fixed-term imprisonment of not more than three years, except as otherwise specifically provided in this Law.

Reading: "China: Case of Assisted Suicide Stirs Euthanasia Debate" (Zeldin 2011, ¶¶1–3). Used with the permission of the *Global Legal Monitor*.

(Aug 17, 2011) In May 2011, the People's Court of Longnan County, in China's Jiangxi Province, sentenced Zhong Yichun, an elderly farmer, to two years' imprisonment for assisting his friend Zeng Qianxiang to commit suicide. Zhong was convicted of criminal negligence resulting in another person's death. (*Farmer Jailed for Assisting Suicide Triggers Controversy*, Xinhua (Aug. 15, 2011), http://www.china.org.cn/china/2011-08/15/content_23210533.htm.)

According to China's Xinhua News Agency, Zeng was mentally ill and had repeatedly asked Zhong to help him commit suicide. In October 2010, Zeng overdosed on sleeping pills and lay down in a hole in the ground; as part of an agreement he had made with Zhong, Zhong buried him after calling out to him 15 minutes later to make sure that Zeng was dead. However, based on an autopsy report indicating that the death was from suffocation, not an overdose, the court found that Zhong failed to confirm Zeng's death before burying him, concluding that Zeng was still alive when Zhong buried him. Zhong appealed the court sentence, but the Intermediate People's Court in Ganzhou City rejected the appeal in August 2011. (*Id.*)

The case has given rise to renewed nation-wide debate on euthanasia, which is banned under current Chinese law; there is no provision permitting assisted suicide. The debate focuses on the question of whether to characterize Zhong's action as intentional homicide or as negligence. (*Id.*) Article 232 of the Criminal Code stipulates a punishment of from three to ten years of fixed-term imprisonment for intentional homicide "if the circumstances are relatively minor"; the penalty is at least ten years of imprisonment up to the death penalty in more serious circumstances. Under article 233, persons who cause another person's death through negligence are punishable on conviction with a prison sentence of from three to seven years; if the circumstances are relatively minor, the maximum sentence is not more than three years, except as otherwise specifically provided in the Code. (Criminal Law of the People's Republic of China, The National People's Congress of the People's Republic of China website [last visited Aug. 17, 2011]).

c. Not a Human Choice

The National Association of Evangelicals (NAE) is composed of a broad array of Christian evangelical organizations including missions, universities, publishers, and churches. Christian evangelicalism represents a growing religious movement in the twenty-first century, one committed to conservative interpretations of biblical teaching. In its 1994 annual conference, the NEA adopted the following statement on physician-assisted suicide or euthanasia: "Human beings are made in the image of God and are, therefore, of inestimable worth. God has given people the highest dignity of all creation. Such human dignity prohibits euthanasia, that is actively causing a person's death" (National Association 1994, ¶1). A 2011 survey of members of the NAE supports a prohibition against physician-assisted suicide based on the view that God would not support such an act (Baklinski 2011). What follows is the resolution adopted at the 1994 annual conference of the NAE.

Reading: "Termination of Medical Treatment" (National Association of Evangelicals 1994)

Human beings are made in the image of God and are, therefore, of inestimable worth. God has given people the highest dignity of all creation. Such human dignity prohibits euthanasia, that is actively causing a person's death.

In the past 30 years, medical technology has developed systems that have enabled physicians to more effectively care for their patients and save lives that would otherwise be lost. However, this technology has also resulted in the possibility of prolonging the dying process beyond its normal course. This often causes great suffering, not only for the patient, but also for the family, friends and caregivers.

Such technology also raises moral questions. For example, is it moral to withdraw a life-support system which is believed to be an inappropriate extension of the dying process? The National Association of Evangelicals (NAE) believes that in cases where patients are terminally ill, death appears imminent and treatment offers no medical hope for a cure, it is morally appropriate to request the withdrawal of life-support systems, allowing natural death to occur. In such cases, every effort should be made to keep the patient free of pain and suffering, with emotional and spiritual support being provided until the patient dies.

When a person's cerebral cortex dies, is it moral for the family or medical staff to withdraw life-support systems? The National Association of Evangelicals believes that in cases where extensive brain injury has occurred and there is clear medical indication that the patient has suffered brain death (permanent unconscious state), no medical treatment can reverse the process.

(Brain death is not the equivalent of a coma. A patient might awaken from a coma, but not from brain death.) Removal of any extraordinary life-support system at this time is morally appropriate and allows the dying process to proceed. Under such circumstances, appropriate action is best taken where there is guidance from a signed "living will" or a durable power of attorney for health care. Where there is no "living will" or durable power of attorney for health care, the decision to withdraw life support should be made by the family and/or closest friends in consultation with a member of the clergy, when available, and the medical staff.

NAE acknowledges that the withdrawal of life-support systems is an emotional and difficult issue. However, we believe that medical treatment that serves only to prolong the dying process has little value. It is better that the dying process be allowed to continue and the patient permitted to die.

This is especially true of those who know Jesus Christ as Savior and Lord. For as the Apostle Paul said: "To be absent from the body is to be present with the Lord" (2 Cor. 5:8).

(Resolution adopted at the 1994 Annual Conference)

The following denominations and fellowships hold membership in the National Association of Evangelicals:

Advent Christian General Conference

Assemblies of God

Baptist General Conference

The Brethren Church (Ashland, Ohio)

Brethren in Christ Church

Christian & Missionary Alliance

Christian Catholic Church (Evangelical Protestant)

Christian Church of North America

Christian Reformed Church in North America

Christian Union

Church of God (Cleveland, Tennessee)

Church of God, Mountain Assembly, Inc.

The Church of the Nazarene

Church of the United Brethren in Christ

Churches of Christ in Christian Union

Conservative Baptist Association

Conservative Congregational Christian Conference

Conservative Lutheran Association

Elim Fellowship

Evangelical Church of North America

Evangelical Congregational Church

Evangelical Free Church of America

Evangelical Friends International of North America

Evangelical Mennonite Church

Evangelical Methodist Church

Evangelical Presbyterian Church

Evangelical Missionary Fellowship

Fellowship of Evangelical Bible Church

Fire Baptized Holiness Church of God of the Americas

Free Methodist Church of North America

General Association of General Baptists

International Church of the Foresquare Gospel
International Pentecostal Church of Christ
International Pentecostal Holiness Church
Mennonite Brethren Churches, USA
Midwest Congregational Christian Fellowship
Missionary Church, Inc.
Open Bible Standard Churches
Pentecostal Church of God
Pentecostal Free Will Baptist Church, Inc.
Presbyterian Church in America
Primitive Methodist Church, USA
Reformed Episcopal Church
Reformed Presbyterian Church of North America
The Salvation Army
Synod of Mid-America (Reformed Church in America)
The Wesleyan Church

C. REFLECTIONS: PHYSICIAN-ASSISTED SUICIDE CONSENT FORM

1.(*) As seen in the Oregon legislation on physician-assisted suicide, much attention is given to the conditions that must be met in order to qualify for physician-assisted suicide. What are the conditions set forth in the Oregon legislation? Do you think these conditions adequately protect patients who seek physician-assisted suicide? If so, why? If not, why not? Develop your reasoning and give details.

2.(*) Do you agree or disagree that death through physician-assisted suicide is morally permissible? Defend your position. Draw from philosophical, clinical, and/or sacred texts to support your view and give details.

3.   This is a continuation of your advance planning in preparation for death: Find the form used to request physician-assisted suicide in the state of Oregon. You can go online and google "Oregon Death with Dignity Act." Read and complete the form so that you become familiar with what patients sign in the state of Oregon when requesting physician-assisted suicide. Note: This is a legal document if one is an Oregonian resident, is terminal, has signed the paperwork, and has a physician willing to prescribe the needed medications. It is not

a legal document if the criteria are not met. Should you not want this to be a legal document, write or type "void" across it after you complete it.

D. FURTHER READINGS

Baklinski, Thaddeus. 2011. "Evangelical Leaders Survey: 'Let God Be God' in End of Life Issues." Lifesitenews.com, February 4. www.lifesitenews.com/news/evangelical-leaders-survey-let-god-be-god-in-end-of-life-issues.

Colby, W. H. 2006. *Unplugged: Reclaiming Our Right to Die in America.* New York: Amacom.

Eastbaugh, Ben, and Chris Sternal-Johnson. 2009. "The Patient Has a Right to Decide When and How to Die!" *Life or Death—Who Decides?* (blog), October 4. http://pulltheplugon life.workpress.com/.

Humphry, Derek. 1986. *The Right to Die.* New York: Harper and Row.

———. 2002. *Final Exit: The Practicalities of Self-Deliverance and Assisted Suicide for the Dying.* 3rd ed. New York: Dell.

Lavi, Shai J. 2005. *The Modern Art of Dying: A History of Euthanasia in the United States.* Princeton, NJ: Princeton University Press.

Lipuma, S. H. 2013. "Continuous Sedation until Death as Physician-Assisted Suicide/ Euthanasia: A Conceptual Analysis." *Journal of Medicine and Philosophy* 23:190–204.

McDonald-Gibson, Charlotte. 2014. "Belgium Extends Euthanasia to Kids." *Time*, February 3. http://time.com/7565/belgium-euthanasia-law-children-assisted-suicide.

Menzel, Paul T., and Bonnie Steinbock. 2013. "Advance Directives, Dementia, and Physician-Assisted Death." *Journal of Law, Medicine and Ethics* 41 (2): 484–500.

Washington Death with Dignity Act. 2013. "Initiative 100." www.doh.wa.gov/YouandYour Family/IllnessandDisease/DeathwithDignityAct.aspx.

Wilkinson, Martin, and Stephen Wilkinson. 2016. "The Donation of Human Organs." In *Stanford Encyclopedia of Philosophy.* http://plato.stanford.edu/entries/organ-donation/.

Young, Robert. 2016. "Voluntary Euthanasia." In *Stanford Encyclopedia of Philosophy.* http://plato.stanford.edu/entries/euthanasia-voluntary/.

Part IV

THE LESSONS
OF DEATH

As this investigation of death comes to a close in part IV: The Lessons of Death, we return to one of the reflection questions posed at the end of chapter 1: what is death? In the previous chapters, we considered a host of responses to this question, including that death is physical disintegration, psychological disintegration, reincarnation, resurrection, medical immortality, digital immortality, an existential phenomenon of life, bad or good, to be feared or not, to be grieved and how, and to be hastened or not in the case of suicide, treatment refusal, and physician-assisted suicide. Such views serve as a basis for a set of exercises that engage readers in reflections about conceptual as well as practical issues concerning the nature of death and that which leads to death.

In thinking about death, one may embrace a number of views on what is death, even ones that may appear to be contradictory to each other. One may hold, for instance, that death is physical disintegration at the same time one may hold that death is reincarnation or resurrection. One may hold that death involves some sense of transcendence (connecting one with one's biological line) while parting with other senses of transcendence (involving an ontological transformation). One may hold that death is both good and bad, to be feared and not to be feared, to be grieved and not to be grieved, and to be accepted and not to be accepted in the case of treatment refusal. One may even hold that, contrary to a dominant religious teaching, physician-assisted suicide may be morally permissible in cases in which an adult patient is terminal and in unrelenting pain. Indeed, our understanding and experience of death can be quite complex and nuanced.

Differences aside, one's view of death carries significant implications for how one lives one's life. It carries implications for how one views the self, how one acts in light of the standards and sanctions of one's belief system, whether one believes in immortality, and whether one judges life to be good or bad, to be feared or not, and to be accepted or not. It carries implications for whether one fears death and approaches death with sadness or acceptance, hope or fate, or heightened awareness or calm neutrality. One's view of death frames one's view of life and how one lives.

One's view of death also carries significant implications for how one dies. It frames how one prepares for death, the kind and degree of medical care one chooses, and the resources and funds that one expends in the dying process. It guides what funeral or memorial service is chosen, what music is selected, what words or prayers are said, what sacraments are given, what foods are served, what clothing is worn, and where and how one is laid to rest. It influences whether one completes a living will, a medical durable power of attorney, or a last will and testament. It marks one's membership within particular traditions of thought and practice and designates points of acceptable and unacceptable actions and rituals.

This exploration of the nature and meaning of death set out to expose readers to a wide variety of views drawn from various times and cultures. Conceptual reflections coupled with practical exercises including planning one's funeral or memorial service give readers an opportunity to think about death in a way that may aid planning for death, whether that is one's own or that of another. In settings that discourage open discussions about death, the exploration offered in this text offers an opportunity to think about death in a setting that is less constrained by time, emotion, and taboo.

This fourth and final part draws our investigation of death to a close with a look at the lessons death, or rather a study of death, imparts to those who take the time for it. A running theme is that death teaches us something about how to live our lives.

Chapter 15

A Window into Life

A. CONTEXT

As French philosopher Michel de Montaigne (1533–92) famously said, echoing the Roman philosopher Cicero (106–43 BCE), "To study philosophy is to learn to die" (1952, 28). Revising this aphorism for our study of death, we could say that "to study philosophy and sacred texts is to learn to live." For Montaigne his statement is true "because study and contemplation do in some sort withdraw from us our soul, and employ it separately from the body, which is a kind of apprenticeship and a resemblance of death; or else, because all the wisdom and reasoning in the world do in the end conclude in this point, to teach us not to fear to die" (1952, 28). In other words, a study of death can help clarify our own views and teach us to rethink cherished views, leading to a death of sorts of some of our taken-for-granted concepts, methods, and values. It can teach us to come to terms with what we know and what we do not or cannot know, thus leading us to rethink the fear we have of death. In short, a study of death through philosophical and sacred texts is an investigation of life and living and what it means to be human.

It does not follow that one needs to agree with all positions found in philosophical and sacred texts. In considering a variety of positions on death, including ones with which one disagrees, one is able in some sense to "get outside" the self to find greater clarity and commitment to the position(s) one embraces. One is also able to see points of conversion and diversion, thus leading one to appreciate our shared humanity in all its diversity. It is in this spirit that this project has been taken: to explore various interpretations and experiences of death for purposes of arriving at some conclusions about death on both the personal and shared levels. While there is no one story of death, some shared themes do appear.

This investigation of the nature and meaning of death comes to a close with selections from a variety of authors about what death can teach us. These include journalist Susie Steiner, who shares her lessons regarding living drawn from her investigation of the wishes of those dying; Rabbi Abraham Joshua Heschel, who reflects on what death teaches us about our participation in the divine; and Tibetan Buddhist lama Rimpoche Nawang Gehlek, who asks his readers to open their minds and see if they can entertain a bigger picture of death and thereby lead their life in a new way.

B. A FINAL PERSPECTIVE

1. An Opportunity to Think about How One Lives

Susan Steiner, writer and editor at the *Guardian*, a British newspaper, penned a report about an Australian nurse, Bronnie Ware, who spent time documenting her dying patients' wishes. In what follows, Steiner shares the story of Ware's findings, including five reported regrets of the dying. One might expect grandiose advice such as "Travel the world" or "Climb Mt. Kilimanjaro." But no, the suggestions are relatively simple, such as "I wish I'd had the courage to live a life true to myself, not the life others expected of me" (Steiner 2012, ¶6) and "I wish that I had let myself be happier" (Steiner 2012, ¶14). Here the dying share lessons with us about how to live our lives.

Reading: "5 Regrets of the Dying" (Steiner 2012). Copyright Guardian News and Media Ltd. 2016. Reprinted with the permission of *The Guardian*.

A nurse has recorded the most common regrets of the dying, and among the top ones is "I wish I hadn't worked so hard." What would your biggest regret be if this was your last day of life?

There was no mention of more sex or bungee jumps. A palliative nurse who has counselled the dying in their last days has revealed the most common regrets we have at the end of our lives. And among the top, from men in particular, is "I wish I hadn't worked so hard."

Bronnie Ware is an Australian nurse who spent several years working in palliative care, caring for patients in the last 12 weeks of their lives. She recorded their dying epiphanies in a blog called Inspiration and Chai, which gathered so much attention that she put her observations into a book called *The Top Five Regrets of the Dying.*

Ware writes of the phenomenal clarity of vision that people gain at the end of their lives, and how we might learn from their wisdom. "When questioned about any regrets they had or anything they would do differently," she says, "common themes surfaced again and again."

Here are the top five regrets of the dying, as witnessed by Ware:

1. I wish I'd had the courage to live a life true to myself, not the life others expected of me.

"This was the most common regret of all. When people realise that their life is almost over and look back clearly on it, it is easy to see how many dreams have gone unfulfilled. Most people had not honoured even a half of their dreams and had to die knowing that it was due to choices they had made, or not made. Health brings a freedom very few realise, until they no longer have it."

2. I wish I hadn't worked so hard.

"This came from every male patient that I nursed. They missed their children's youth and their partner's companionship. Women also spoke of this regret, but as most were from an older generation, many of the female patients had not been breadwinners. All of the men I nursed deeply regretted spending so much of their lives on the treadmill of a work existence."

3. I wish I'd had the courage to express my feelings.

"Many people suppressed their feelings in order to keep peace with others. As a result, they settled for a mediocre existence and never became who they were truly capable of becoming. Many developed illnesses relating to the bitterness and resentment they carried as a result."

4. I wish I had stayed in touch with my friends.

"Often they would not truly realise the full benefits of old friends until their dying weeks and it was not always possible to track them down. Many had become so caught up in their own lives that they had let golden friendships slip by over the years. There were many deep regrets about not giving friendships the time and effort that they deserved. Everyone misses their friends when they are dying."

5. I wish that I had let myself be happier.

"This is a surprisingly common one. Many did not realise until the end that happiness is a choice. They had stayed stuck in old patterns and habits. The so-called 'comfort' of familiarity overflowed into their emotions, as well as their physical lives. Fear of change had them pretending to others, and to their selves, that they were content, when deep within, they longed to laugh properly and have silliness in their life again."

2. A Way to Understand the Meaning of Life

Rabbi Abraham Joshua Heschel (1907–72) is a prominent Jewish spiritual teacher and philosopher of the twentieth century. In the reading that follows, Heschel argues that facing death gives life meaning. One thing that death teaches us is that there is more to life than simple bodily existence. Life and death are both part of a greater mystery and humans are created in no less than God's image. "Man partakes of an unearthly divine sort of being" (Heschel 1974, 61). We know this because we can imagine an afterlife for humanity. As Heschel says, "Indeed, man's hope for eternal life presupposes that there is something about man that is worthy of eternity, that has some affinity to what is divine, that is made in the likeness of the divine" (Heschel 1974, 59). At the same time, death itself is an antidote to human arrogance. In death, we pay gratitude for the wonder and gift of our existence. The following reading appears in Heschel's essay "Death as Homecoming," published in *Jewish Reflections on Death*, edited by Jack Riemer.

Reading: "Death as Homecoming" (Heschel 1974, 58–59, 61). Copyright Professor Susannah Heschel, reprinted with permission.

Our first question is to what end and upon what right do we think about the strange and totally inaccessible subject of death? The answer is because of the supreme certainty we have about the existence of man: that it cannot endure without a sense of meaning. But existence embraces both life and death, and in a way death is the test of the meaning of life. If death is devoid of meaning, then life is absurd. Life's ultimate meaning remains obscure unless it is reflected upon in the face of death.

The fact of dying must be a major factor in our understanding of living. Yet only few of us have come face to face with death as a problem or a challenge. There is a slowness, a delay, a neglect on our part to think about it. For the subject is not exciting, but rather strange and shocking.

What characterizes modern man's attitude toward death is escapism, disregard of its harsh reality, even a tendency to obliterate grief. He is entering, however, a new age of search for meaning of existence, and all cardinal issues will have to be faced.

Death is grim, harsh, cruel, a source of infinite grief. Our first reaction is consternation. We are stunned and distraught. Slowly, our sense of dismay is followed by a sense of mystery. Suddenly, a whole life has veiled itself in secrecy.

Our speech stops, our understanding fails. In the presence of death there is only silence, and a sense of awe.

Is death nothing but an obliteration, an absolute negation? The view of death is affected by our understanding of life. If life is sensed as a surprise, as a gift, defying explanation, then death ceases to be a radical, absolute negation of what life stands for. For both life and death are aspects of a greater mystery, the mystery of being, the mystery of creation. Over and above the preciousness of particular existence stands the marvel of its being related to the infinite mystery of being or creation.

Death, then, is not simply man's coming to an end. It is also entering a beginning.

There is, furthermore, the mystery of my personal existence. The problem of how and whether I am going to be after I die is profoundly related to the problem of who and how I was before I was born. The mystery of an afterlife is related to the mystery of preexistence. A soul does not grow out of nothing. Does it, then, perish and dissolve in nothing?

Human life is on its way from a great distance; it has gone through ages of experience, of growing, suffering, insight, action. We are what we are by what we come from. There is a vast continuum preceding individual existence, and it is a legitimate surmise to assume that there is a continuum following individual existence. Human living is always being under way, and death is not the final destination.

In the language of the Bible to die, to be buried, is said to be "gathered to his people" (Genesis 25:8). They "were gathered to their fathers" (Judges 2:10). "When your days are fulfilled to go to be with your fathers" (I Chronicles 17:11).

Do souls become dust? Does spirit turn to ashes? How can souls, capable of creating immortal words, immortal works of thought and art, be completely dissolved, vanish forever?

Others may counter: The belief that man may have a share in eternal life is not only beyond proof; it is even presumptuous. Who could seriously maintain that members of the human species, a class of mammals, will attain eternity? What image of humanity is presupposed by the belief in immortality?

Indeed, man's hope for eternal life presupposes that there is something about man that is worthy of eternity, that has some affinity to what is divine, that is made in the likeness of the divine. . . .

Thus, the likeness of God means the likeness of Him who is unlike man. The likeness of God means the likeness of Him compared with whom all else is like nothing.

Indeed, the words "image and likeness of God" [in the biblical Creation story] conceal more than they reveal. They signify something which we can neither comprehend nor verify. For what is our image? What is our likeness? Is there anything about man that may be compared with God? Our eyes do not see it; our minds cannot grasp it. Taken literally, these words are absurd, if not blasphemous. And still they hold the most important truth about the meaning of man.

Obscure as the meaning of these terms is, they undoubtedly denote something *unearthly,* something that belongs to the sphere of God. *Demut* [likeness] and *tzelem* [image] are of a higher sort of being than the things created in the six days. This, it seems, is what the verse intends to convey: Man partakes of an unearthly divine sort of being.

3. A Lesson on Leading a Good Life

Rimpoche Nawang Gehlek (1939–present) is a Tibetan Buddhist lama or monk born in Lhasa, Tibet. He is nephew of the thirteenth Dalai Lama. In 1959, he fled to India from Tibet after the communist Chinese invasion, socialist land reform, and military crackdown and gave up monastic life. Since then, he has devoted his life to teaching about Buddhism. In his *Good Life, Good Death,* Rimpoche Nawang Gehlek addresses the question "How does our understanding of death affect how we live our lives?" Here is part of his answer:

> Death is definite, as we know. No one can avoid it. Even spiritually developed people such as Buddha could not avoid it. No one has ever lived forever. No one reading of this book will live forever, no matter how old, young, beautiful, rich, poor, or highly developed that person might be. But instead of running away from the thought of death, we have a better chance of doing something for ourselves if we take a look at what's coming or at least try to imagine it. Not only will this help reduce our fear, it will lay the groundwork for us to take advantage of the opportunity to transform the dying process into a process of enlightenment. (Gehlek 2001, 145)

In what follows, Rimpoche Nawang Gehlek asks his readers to open their minds and see if they can entertain a bigger picture of death and thereby lead life in a new and fuller way.

Death is definite, as we know. No one can avoid it. Even spiritually developed people such as Buddha could not avoid it. No one has ever lived forever. No one reading of this book will live forever, no matter how old, young, beautiful, rich, poor, or highly developed that person might be. But instead of running away from the thought of death, we have a better chance of doing something for ourselves if we take a look at what's coming or at least try to imagine it. Not only will this help reduce our fear, it will lay the groundwork for us to take advantage of the opportunity to transform the dying process into a process of enlightenment. And if we can't accomplish that, at least we can have a better death.

Whatever life force we have was there from the beginning. There is no extension possible, and the days, weeks, months, and years are drawn from the original supply we were given. One day, it will be like a dry pond after all the water has evaporated. Our conditions for living can easily become causes for dying. The chemicals in our bodies deteriorate. The wrong food or the wrong medicine can have a bad side effect.

We should accept this now. If I wait until I'm actually dying, it will be too late to do anything. So I must accept that I am definitely going to go and that death is definitely going to come for me. Nobody knows when. Since nobody knows when, it could be next week, next month, or next year. I have no certainty that I'll even be here tomorrow or an hour from now. So if I understand this to be so, I must resolve to use whatever time I have to put my anger, attachment, and ego to rest. When I die, when my consciousness leaves my body, what can I take with me other than the karmic imprint of good and bad; virtue or non-virtue; positive or negative? I'll need plenty of positive imprints. In fact, I want only positive ones. But if that's not possible, at least I want to be able to connect with my good fortune before I connect with any negative karma I may have accumulated.

Though death is nothing more than the separation of the body that we have used in one life from the mind that accompanies us into every life, strong emotions are bound to surface at the time of death. Death is the end of all our activities in this life, good and bad. We'll suffer at the thought of not seeing or feeling anything anymore, of not being able to be with or talk to loved ones. Not letting go is our biggest problem. Say what you have to say to the people who matter to you, or write what you have to write. But beyond that, hanging on to resentments or to intense attachment is not good, either for the person dying or for those left behind. It's important to use your understanding and your willpower to cut bad feelings, and if necessary, cut them drastically.

C. REFLECTIONS: WHAT IS DEATH?

1.(*) As you close your reflections, what in your view is death? Have your views of death and dying changed after your reflections?

2.(*) Interview someone else about his or her views of death and dying. Summarize what you hear in your conversation.

3. If you have recommendations for additional topics or materials to be included in this text, please share them with me. This text is a product of reader feedback over the last many years, and I hope it continues to be a fluid project. Thanks for all of your reflections. May you live—and die—well!

D. FURTHER READINGS

Albom, Mitch. 1997. *Tuesdays with Morrie.* New York: Doubleday.

Derrida, Jacques. 1986. *Memories for Paul de Man.* Translated by C. Lindsay et al. New York: Columbia University Press.

Gawande, Atul. 2014. *Being Mortal: Medicine and What Matters in the End.* New York: Metropolitan Books.

Klein, Ezra. 2015. "9 Lessons a Physician Learned about Dying." *Vox*, March 26. www.vox.com/2014/10/21/7023257/atul-gawande-taught-me-dying-being-mortal.

Kushner, Harold. 1978. *When Bad Things Happen to Good People.* New York: Random House.

Nuland, Sherwin. 1995. *How We Die: Reflections on Life's Final Chapter.* New York: Vintage.

Pausch, Randy. 2008. *The Last Lecture.* www.thelastlecture.com/.

Quindlen, Anna. 2000. *A Short Guide to a Happy Life.* New York: Random.

Ware, Bronnie. 2012. *The Top Five Regrets of the Dying: A Life Transformed by the Dearly Departed.* New York: Hay House.

— Glossary of Philosophical Terms —

absurd: In the existential tradition, that which is meaningless.

anguish: In the existential tradition, the reflective apprehension of the self and the realization that nothing relieves me from choosing myself.

autonomy: The ethical principle of self-determination. Autonomy involves two conditions: (1) liberty, or independence from controlling influences; and (2) agency, capacity for intentional action.

axiology: The philosophical study of value.

beneficence: The ethical principle of advancing another's interest. Beneficence entails various kinds of duties, such as the protection and defense of the rights of others, the prevention of harm from occurring to others, the removal of conditions that will cause harm to others, help for persons with disabilities, and the rescue of persons who are in danger.

bioethics/biomedical ethics: The study of ethical issues in biomedicine. Bioethics/biomedical ethics emerged in the twentieth century in reaction to the atrocities that occurred in World War II.

body: The physical manifestation of an animal or plant.

brain death: End of brain functioning. There are whole-brain and higher-brain definitions of death.

consciousness: A term that refers to awareness or self-reflection.

contextualism: The view in philosophy that claims depend on the context.

death: The end of life. Otherwise, definitions vary in terms of time, place, culture, religion, and spirituality.

digital immortality: Life after death made possible by digital technology.

empiricism: The view in philosophy that knowledge is derived from sense perception.

enlightenment: In Asian philosophy, refers to liberation or freedom from suffering.

epistemology: The philosophical study of knowledge.

ethical: That which is moral, good, right, or virtuous.

ethics: The philosophical study of good and bad, and right and wrong, as these judgments have to do with the actions and character of individuals, families, communities, institutions, and society.

existentialism: The view in philosophy that emphasizes living life through individual acts of the will.

fact: Something that is said to exist or has happened.

freedom: In the existential tradition, the very being for-itself, which is "condemned to be free" and must choose itself, i.e., make itself.

higher-brain definition of death: End of functioning of the cortical area of the brain.

immortality: State of being subject to life after death.

informed consent: In medicine, the process of obtaining permission from a patient before an intervention occurs. Involves disclosure of medical information, options, and consequences.

liberation: Enlightenment or freedom from suffering.

materialism: The view in philosophy that reality is composed of matter or that which is physical. See "physicalism."

medical immortality: Immortality or life after death made possible by medical technology.

metaphysics: The philosophical study of that which is beyond the physical. In phenomenology, the study of individual processes that have given birth to this world as a concrete and particular totality.

moral: That which is ethical, or good, right, or virtuous.

moral value: Sign of praiseworthiness of blameworthiness.

mortality: State of being subject to death.

naturalism: The view in philosophy that reality is composed of natural forces and entities.

nonmaleficence: The ethical principle of doing no harm.

normativism: The view in philosophy that norms or values frame claims of fact or understanding.

objectivity: The state of being independent of mind. Usually contrasts with subjectivity.

ontology: The philosophical study of being.

physical disintegration: The process that leads toward death of the body.

physicalism: The view in philosophy that reality is physical or material. See "materialism."

possible: In existentialism, an action to be performed in a real world, rather than an abstract idea of possibility in general.

psychological disintegration: The process that leads toward death of consciousness.

rationalism: The view in philosophy that knowledge is derived from reason.

realism: The view in philosophy that reality is independent of human perception or thought.

reason: A basis or cause of belief, action, fact, event, or sentiment.

reductionism: The view in philosophy that complex phenomena are explainable in terms of their parts.

relativism: The view in philosophy that claims are relative to cultural and social standards.

right: In philosophy and law, a moral or legal entitlement.

sacred texts: Writings that are revered by a community of believers.

skepticism: The view in philosophy that claims are to be doubted.

soul: The nonphysical self, essence, or "I" that inhabits a body.

Stoicism: The view in philosophy that emphasizes the acceptance of life's hardships.

subjectivity: The state of being dependent on the mind. Compare with "objectivity."

value: A sign of worth or assessment of comparison. In philosophy, a value can be nonmoral (as in the case of money) or moral (as in the case of respect for autonomy).

whole-brain definition of death: End of brain functioning that comes when entire brain ceases to function.

— References —

Ad Hoc Committee of the Harvard Medical School. 1968. "A Definition of Irreversible Coma: Report of the Ad Hoc Committee of the Harvard Medical School to Examine the Definition of Brain Death." *Journal of the American Medical Association* 205 (6): 337–40.

Albom, Mitch. 1997. *Tuesdays with Morrie.* New York: Doubleday.

Alexander, Eben. 2012. *Proof of Heaven: A Neurosurgeon's Journey into the Afterlife.* New York: Simon and Schuster.

Allen, Kelley. 2013. "Eastern Traditions versus Western Beliefs as Related to the Grief Process Regarding End-of-Life Care in the Neonatal Intensive Care Unit." www.national perinatal.org/index/pdf/317B-paper.pdf. No longer available online.

Andrade, Gabriel. 2017. "Immortality." In *Internet Encyclopedia of Philosophy.* https://www .iep.utm.edu/immortal/.

Annas, George J. 2008. "Immortality through Cloning." In *Aging, Biotechnology, and the Future*, edited by Catherine Y. Read, Robert C. Green, and Michael A. Smyer, 17–38. Baltimore: Johns Hopkins University Press.

Aquinas, Thomas. 1947. *Summa Theologica.* Benzinger Bros. ed. Translated by Fathers of the English Dominican Province. www.ccel.org/a/aquinas/summa/home.html.

Aries, Philippe. 1975. *Western Attitudes toward Death: From the Middle Ages to the Present.* Baltimore: Johns Hopkins University Press.

Aristotle. 1984. *On Youth, Old Age, Life and Death, and Respiration.* In *The Complete Works of Aristotle: The Revised Oxford Translation*, vol. 1, edited by Jonathan Barnes, 751–61. Princeton, NJ: Princeton University Press.

"Aristotle." 2017. *Wikipedia,* n.d., accessed July 6. http://en.wikipedia.org/wiki/Aristotle.

Augustine. 2000. *The City of God.* Translated by Marcus Dods. New York: Modern Library.

Aurelius, Marcus. 1909. *The Meditations of Marcus Aurelius.* Translated by George Long. New York: P. F. Collier and Sons. http://classics.mit.edu/Antoninus/meditations.html.

Avesta: Yasna: Sacred Liturgy and Gathas/Hymns of Zarathushtra. [1898] 1995. Translated by L. H. Mills. Sacred Books of the East. Digital copyright by Joseph H. Peterson. http: //avesta.org/yasna/yasna.htm.

Baklinski, Thaddeus. 2011. "Evangelical Leaders Survey: 'Let God Be God' in End of Life Issues." Lifesitenews.com, February 4. www.lifesitenews.com/news/evangelical-leaders-survey-let-god-be-god-in-end-of-life-issues.

Barnes, Hazel E. 1959. *The Literature of Possibility: A Study in Humanistic Existentialism.* Lincoln: University of Nebraska Press.

Barrow, John, and Frank Tipler. 1988. *The Anthropic Cosmological Principle.* New York: Oxford University Press.

Barry, Vincent. 2007. *Philosophical Thinking about Death and Dying.* Belmont, CA: Thomson Wadsworth.

Beauchamp, Tom L., and James F. Childress. 2013. *Principles of Biomedical Ethics.* 7th ed. New York: Oxford University Press.

Beauvoir, Simone de. [1949] 1989. *The Second Sex.* Trans. H. M. Parshley. New York: Vintage Books.

———. 1985. *A Very Easy Death.* New York: Pantheon Books.

Becker, Ernest. 1973. *The Denial of Death.* New York: Free Press Paperbacks.

———. 1997. *The Fear of Death.* New York: First Press Perspectives.

Bell, Gordon, and Jim Gray. 2001. "Digital Immortality." *Communications of the ACM* 44 (3): 29–31. https://gordonbell.azurewebsites.net/CGB%20Files/CACM%20Digital%20Immortality%20with%20Gray%200103%20c.pdf.

Benecke, Mark. 2002. *The Dream of Eternal Life: Biomedicine, Aging, and Immortality.* New York: Columbia University Press.

Berlinger, Nancy, Bruce Jennings, and Susan M. Wolf. 2013. *The Hastings Center Guidelines for Decisions on Life-Sustaining Treatment and Care near the End of Life.* 2nd ed. New York: Oxford University Press.

Bhagavad-Gita. 1996. In *Hindu Scriptures*, edited and translated by Dominic Goodall. Berkeley: University of California Press.

Bhattacharya, Pranab Kumar. 2013. "Is There Science behind the Near-Death Experience: Does Human Consciousness Survive after Death?" *Annals of Tropical Medicine and Public Health* 6 (2): 151–65.

Bilhartz, Terry D. 2006. *Sacred Words: A Source Book on the Great Religions of the World.* New York: McGraw Hill.

Bonanno, George A. 2009. *The Other Side of Sadness: What the New Science of Bereavement Tells Us about Life after Loss.* New York: Basic Books.

Bova, Ben. 1998. *Immortality: How Science Is Extending Your Life Span—and Changing the World.* New York: Avon Books.

Brennan, Samantha, and Robert J. Stainton, eds. 2009. *Philosophy and Death: Introductory Readings.* Toronto, Ontario: Broadview Press.

Buben, Adam, and Patrick Stokes. 2011. *Kierkegaard and Death.* Bloomington: Indiana University Press.

Bunzel, Ruth L. 1929–30. *Zuñi Ritual Poetry.* Forty-Seventh Annual Report of the Bureau of American Ethnology. Washington, D.C.: Smithsonian Institution.

Byrock, Ira. 1998. *Dying Well.* New York: Riverhead.

————. 2012. *The Best Care Possible.* New York: Avery.

Callanan, Maggie, and Patricia Kelley. [1992] 2012. *Final Gifts: Understanding the Special Awareness, Needs, and Communications of the Dying.* New York: Simon and Schuster.

Camus, Albert. 1955. *The Myth of Sisyphus and Other Essays.* Translated by Justin O'Brien. New York: Alfred A. Knopf.

Canadian Congress Committee on Brain Death. 1988. "Death and Brain Death: A New Formulation for Canadian Medicine." *Canadian Medical Association Journal* 138: 405–6.

Capps, Walter H. 1995. *Religious Studies: The Making of a Discipline.* Minneapolis: Augsburg Fortress.

Carr, Thomas K. 2006. *Introducing Death and Dying: Readings and Exercises.* Upper Saddle River, NJ: Pearson/Prentice-Hall.

Carter, Chris. 2010. *Science and the Near-Death Experience: How Consciousness Survives Death.* Rochester, VT: Inner Traditions.

Catechism of the Catholic Church: Modifications from the Editio Typica. 1997. Washington, D.C.: United States Catholic Conference. www.usccb.org/beliefs-and-teachings/what-we-believe/catechism/catechism-of-the-catholic-church/epub/index.cfm.

Cave, Stephen. 2012. *Immortality: The Quest to Live Forever and How It Drives Civilization.* New York: Crown.

Chalmers, David. 2002. *Philosophy of Mind: Classical and Contemporary Readings.* Oxford: Oxford University Press.

Chaudhuri, Saabira. 2011. "The 25 Documents You Need before You Die." *Wall Street Journal,* July 2. http://www.wsj.com/articles/SB10001424052702303627104576410234039258092.

Chidester, David. 2002. *Patterns of Transcendence: Religion, Death, and Dying.* Belmont, CA: Wadsworth.

Cholbi, Michael. 2017. "Suicide." In *Stanford Encyclopedia of Philosophy.* http://plato.stanford.edu/entries/suicide/.

"Chris Faraone." 2017. *Wikipedia,* n.d., accessed July 2. http://en.wikipedia.org/wiki/Chris_Faraone.

Chu, Leung Wing. 2012. "One Step Forward for Advance Directives in Hong Kong." *Hong Kong Medical Journal* 18 (3): 176–77. www.hkmj.org/system/files/hkm1206p176.pdf.

Chu, Leung Wing, James K. H. Luk, Elsie Hui, Patrick K. C. Chiu, Cherry S. Y. Chan, Fiona Kwan, Timothy Kwok, Diana Lee, and Jean Woo. 2011. "Advance Directive and End-of-Life Care Preferences among Chinese Nursing Home Residents in Hong Kong." *Journal of the American Medical Directors Association* 12 (2): 143–52. http://hub.hku.hk/bitstream/10722/139467/1/Content.pdf?accept=1.

Chuang Tzu. 1964. "Supreme Happiness." In *Chuang Tzu: Basic Writings,* translated by Burton Watson, 113–15. New York: Columbia University Press.

Colby, W. H. 2006. *Unplugged: Reclaiming Our Right to Die in America.* New York: Amacom.

Colorado Medical Durable Power of Attorney. 2012. Title 15, C.R.S., 15-14-506. www.lexis
	nexis.com/hottopics/colorado/.

Colorado Medical Treatment Decision Act. 2012. Title 15, C.R.S. 15-18-104. www.lexis
	nexis.com/hottopics/colorado/.

Conte, H. R., M. B. Weiner, and R. Plutchik. 1982. "Measuring Death Anxiety: Conceptual,
	Psychometric, and Factor-Analytic Aspects." *Journal of Personality and Social Psy-
	chology* 43 (4): 775–85.

Criminal Law of the People's Republic of China. 2013. Articles 232 and 233. www.npc
	.gov.cn/englishnpc/Law/2007-12/13/content_1384075.htm.

Critchley, Simon. 2008. *The Book of Dead Philosophers.* New York: Vintage Books.

DeGrazia, David. 1998. "Biology, Consciousness, and the Definition of Death." *Report from
	the Institute of Philosophy and Public Policy* 18 (1–2): 18–22. https://www.researchgate
	.net/publication/11695025_Biology_consciousness_and_the_definition_of_death.

———. 2017. "The Definition of Death." In *Stanford Encyclopedia of Philosophy.* http:
	//plato.stanford.edu/entries/death-definition/.

DeMaria, Anthony N. 2010. "Problems with Immortality." *Journal of the American College of
	Cardiology* 56 (25): 2140–42. http://content.onlinejacc.org/article.aspx?articleid=1144
	004.

Derrida, Jacques. 1986. *Memories for Paul de Man.* Translated by C. Lindsay et al. New
	York: Columbia University Press.

Descartes, René. [1641] 2009. *Discourse on Method.* In *Modern Philosophy,* 2nd ed., edited
	by Roger Ariew and Eric Walters, 137–42. Indianapolis: Hackett.

———. [1649] 1985. *Passions of the Soul.* In *The Philosophical Writings of Descartes,* vol. 1,
	translated by Robert Stoothoff, 328–404. Cambridge: Cambridge University Press.

Despelder, Lynne Ann, and Albert Lee Strickland. 2014. *The Last Dance: Encountering
	Death.* 10th ed. New York: McGraw Hill.

Devine, Claire. 2010. "Tissue Rights and Ownership: Is a Cell Clone a Research Tool or a
	Person?" *Columbia Science and Technology Law Review* 20 (March 9). www.stlr.org
	/2010/03/tissue-rights-and-ownership-is-a-cell-line-a-research-tool-or-a-person/.

Diop, Birago. 1991. "The Dead Are Never Gone." In *World Scripture: A Comparative Anthol-
	ogy of Sacred Texts,* edited by Andrew Wilson, 232–33. St. Paul, MN: Paragon House.

———. 2000. *Leurres et lueurs: Poèmes.* 3rd ed. Paris: Présence Africaine.

Dixon, R. M. W. 1996. "Origin Legends and Linguistic Relationships." *Oceania* 67 (2):
	127–40.

Doka, Kenneth J., and Terry L. Martin. 2010. *Grieving beyond Gender: Understanding the
	Ways Men and Women Mourn.* New York: Taylor and Francis.

Durkheim, Emile. 1951. *On Suicide: A Study in Sociology.* New York: Free Press.

Eastbaugh, Ben, and Chris Sternal-Johnson. 2009. "The Patient Has a Right to Decide
	When and How to Die!" *Life or Death—Who Decides?* (blog), October 4. http://pullthe
	plugonlife.workpress.com/.

Edwards, Paul, ed. 1997. *Immortality.* New York: Prometheus Books.

———. 2001. *Reincarnation: A Critical Examination.* New York: Prometheus Books.

The Egyptian Book of the Dead. Edited by J. Romer. 2008. Translated by E. A. Wallis Budge. New York: Penguin Books.

Elisabeth Kübler-Ross Foundation. 2015. "EKR Biography." http://www.ekrfoundation .org/bio.

Encyclopedia of Death and Dying. 2017. www.deathreference.com.

Engelhardt, H. Tristram, Jr. 1996. *The Foundations of Bioethics.* 2nd ed. New York: Oxford University Press.

The Epic of Gilgamesh. 1972. Edited and translated by N. K. Sandars. London: Penguin.

Epictetus. 1983. *The Handbook.* Translated by Nicholas White. Indianapolis: Hackett.

Epicurus. 1963. "Letter to Menoeceus." In *The Philosophy of Epicurus,* translated by George K. Strodach, 179–81. Evanston, IL: Northwestern University Press.

"Epicurus." 2017. *Wikipedia,* n.d., accessed June 28. http://en.wikipedia.org/wiki/Epicurus.

"Euthanasia and Physician-Assisted Suicide." 2016. *Religious Tolerance.* www.religioustoler ance.org/euthanas.htm#.

Faraone, Chris. 2011. "Digital Death: Where Does Your Data Go When You Reach the End of the Road?" *Colorado Springs Independent,* July 28–August 3, 17.

Faure, G. 2009. "How to Manage Your Online Life When You're Dead." *Time,* August 18. www.time.com/time/business/article/0,8599,1916317,00.html.

Feifel, Herman. 1959. *The Meaning of Death.* New York: McGraw-Hill.

Feldmann, Susan, ed. 1963. *African Myths and Tales.* New York: Dell.

Fingarette, Herbert. 1996. *Death: Philosophical Soundings.* Chicago: Open Court.

Fletcher, D. 2009. "What Happens to Your Facebook after You Die?" *Time,* October 28. www.time.com/time/business/article/0,8599,1932803,00.html.

Foley, Elizabeth. 2011. *The Law of Life and Death.* Cambridge, MA: Harvard University Press.

Forsloff, Carol. 2009. "Religion and the Refusal of Medical Treatment: Rights and Responsibilities—Adults Can Say No for Themselves but Not for Children." *Yahoo! Voices,* February 18. http://voices.yahoo.com/religion-refusal-medical-treatment-rights-264 1729.html.

Gawande, Atul. 2014. *Being Mortal: Medicine and What Matters in the End.* New York: Metropolitan Books.

Gehlek, Rimpoche Nawang. 2001. "A Good Death." In *Good Life, Good Death: Tibetan Wisdom on Reincarnation,* 145–47. New York: Riverhead Books.

Goodall, Dominic. 1996. *Hindu Scriptures.* Edited and translated by Dominic Goodall. Berkeley: University of California Press.

Goswami, Amit. 2001. *Physics of the Soul: The Quantum Book of Living, Dying, Reincarnation, and Immortality.* Charlottesville, VA: Hampton Roads.

Hanh, Thich Nhat. 2003. *No Death, No Fear: Comforting Wisdom for Life.* New York: Riverhead Books.

Harris, John. 2000. "Intimations of Immortality." *Science* 288 (5463): 59.

———. 2007. *Enhancing Evolution: The Ethical Case for Making People Better.* Princeton, NJ: Princeton University Press.

Harvey, Peter. 1990. *An Introduction to Buddhism.* Cambridge: Cambridge University Press.

Hasker, William, and Charles Taliaferro. 2017. "Afterlife." In *Stanford Encyclopedia of Philosophy.* http://plato.stanford.edu/entries/afterlife/.

Hayasaki, Erika. 2014. *The Death Class: A Story about Life.* New York: Simon and Schuster.

The Hebrew Bible in English. [1917] 2002. Translated by the Jewish Publication Society. Jerusalem: Mechon Mamre. www.mechon-mamre.org/e/et/et0.htm.

Heidegger, Martin. [1927] 1996. *Being and Time: A Translation of Sein und Zeit.* Translated by Joan Stambaugh. New York: SUNY Press.

Heron, Melonie. 2012. "Deaths: Leading Causes for 2009." *National Vital Statistics Reports* 61 (7): 1–96.

Heschel, Abraham Joshua. 1974. "Death as Homecoming." In *Jewish Reflections on Death*, edited by Jack Riemer, 58–73. New York: Schocken Books.

High, Dallas. 2003. "Death: Philosophical and Theological Foundations." In *Encyclopedia of Bioethics*, 3rd ed., edited by S. G. Post, vol. 1, 301–7. New York: Macmillan.

Homer. 1937. *The Odyssey.* Translated by W. H. D. Rouse. New York: Signet Classic.

"Homer." 2017. *Wikipedia*, n.d., accessed July 23. http://en.wikipedia.org/wiki/Homer.

Hume, David. 1777. "Of Suicide." In *Two Essays.* Edited by Amyas Merivale. London. www.davidhume.org/texts/suis.html.

Humphry, Derek. 1986. *The Right to Die.* New York: Harper and Row.

———. 2002. *Final Exit: The Practicalities of Self-Deliverance and Assisted Suicide for the Dying.* 3rd ed. New York: Dell.

"Immanuel Kant." 2017. *Wikipedia*, n.d., accessed July 20. http://en.wikipedia.org/wiki/Immanuel_Kant.

Jahn, Janheinz. 1961. *Muntu: An Outline of the New African Culture.* Translated by Marjorie Grene. New York: Grove Press.

Jaworski, Katrina. 2010. "The Gender-ing of Suicide." *Australian Feminist Studies* 25 (63): 47–61.

Jones, Richard A. 2016. "The Technology of Immortality, the Soul, and Human Identity." In *Postmodern Racial Dialectics: Philosophy beyond the Pale*, chap. 8, 179–96. Lanham, MD: University Press of America.

Jonsen, Albert R. 1998. *The Birth of Bioethics.* New York: Oxford University Press, 1998.

Kagan, Shelly. 2012. *Death.* New Haven, CT: Yale University Press.

Kant, Immanuel. [1785] 2012. *Groundwork of the Metaphysics of Morals.* Translated by T. K. Abbott. Wikisource, The Free Library. http://en.wikisource.org/wiki/Groundwork_of_the_Metaphysics_of_Morals.

Kastenbaum, Robert. 2016. *Death, Society, and Human Experience.* 11th ed. New York: Pearson Education.

Katha Upanishad. 1996. In *Hindu Scriptures*, edited and translated by Dominic Goodall, 178–80. Berkeley: University of California Press.

Kaufman, Walter. 1989. *Existentialism from Dostoyevsky to Sartre.* New York: Meridian.

Kessler, David. 2007. *The Needs of the Dying: A Guide for Bringing Hope, Comfort, and Love to Life's Final Chapter*. New York: Harper Perennial.

Kierkegaard, Søren. [1849] 1954. *Fear and Trembling and The Sickness unto Death*. Translated by Walter Lowie. Princeton, NJ: Princeton University Press.

Klein, Ezra. 2015. "9 Lessons a Physician Learned about Dying." *Vox*, March 26. www.vox.com/2014/10/21/7023257/atul-gawande-taught-me-dying-being-mortal.

The Kojiki. 1919. Translated by Basil Hall Chamberlain. www.sacred-texts.com/shi/kj/kj016.htm.

Konigsberg, Ruth Davis. 2011. *The Truth about Grief: The Myth of Its Five Stages and the New Science of Loss*. New York: Simon and Schuster.

Kramer, Kenneth. 1988. *The Sacred Art of Dying*. Mahwah, NJ: Paulist Press.

Kramer, Scott, and Kuang-Mong Wu. 1988. *Thinking through Death*. 2 vols. Malabar, FL: Robert E. Krieger.

Kübler-Ross, Elisabeth. 1969. *On Death and Dying: What the Dying Have to Teach Doctors, Nurses, Clergy, and Their Own Families*. New York: Simon and Schuster.

Kumar, Sameet M. 2005. *Grieving Mindfully: Compassionate and Spiritual Guide to Coping with Loss*. Oakland, CA: New Harbinger Publications.

Kushner, Harold. 1978. *When Bad Things Happen to Good People*. New York: Random House.

Kwak, Jung, and William E. Haley. 2005. "Current Research Findings on End-of-Life Decision Making among Racially or Ethnically Diverse Groups." *Gerontologist* 45 (5): 634–41. www.ncbi.nlm.nih.gov/pubmed/16199398.

Lao-Tzu. 1891. *Tao Te Ching*. Translated by J. Legge. Sacred Books of the East 39. www.sacred-texts.com/tao/taote.htm.

Lavi, Shai J. 2005. *The Modern Art of Dying: A History of Euthanasia in the United States*. Princeton, NJ: Princeton University Press.

Lewis, C. S. 1961. *A Grief Observed*. New York: HarperOne.

Lipuma, S. H. 2013. "Continuous Sedation until Death as Physician-Assisted Suicide/Euthanasia: A Conceptual Analysis." *Journal of Medicine and Philosophy* 38: 190–204.

Lizza, John. 2006. *Persons, Humanity, and the Definition of Death*. Baltimore: Johns Hopkins University Press.

Locke, John. [1689] 1980. *Second Treatise of Government*. Edited by C. B. Macpherson. Indianapolis: Hackett.

———. 1959. "Personal Identity." In *An Essay Concerning Human Understanding*, edited by A. C. Fraser, 448–52. New York: Dover Publications.

Lommel, Pim van. 2010. *Consciousness beyond Life: The Science of the Near-Death Experience*. New York: HarperOne.

Luper, Stephen. 2009. *The Philosophy of Death*. Cambridge: Cambridge University Press.

———. 2017. "Death." In *Stanford Encyclopedia of Philosophy*. http://plato.stanford.edu/entries/death.

Lyotard, Jean-François. 1984. *The Postmodern Condition: A Report on Knowledge*. Translated by Geoffrey Bennington and Brian Massumi. Minneapolis: University of Minnesota Press.

Madigen, Kevin J., and Jon D. Levenson. 2008. *Resurrection: The Power of God for Christians and Jews.* New Haven, CT: Yale University Press.

Malpas, Jeff E., and Robert C. Solomon. 1998. *Death and Philosophy.* London: Routledge.

McDonald-Gibson, Charlotte. 2014. "Belgium Extends Euthanasia to Kids." *Time,* February 3. http://time.com/7565/belgium-euthanasia-law-children-assisted-suicide.

McNealy, Kristie. 2010. "In Pursuit of Immortality: The Science behind Life Extension." *Family Health Guide,* July 13, 1–2. www.familyhealthguide.co.uk/in-pursuit-of-immortality-the-science behind-life-extension.html.

McQuire, Meredith. 2008. *Religion: The Social Context.* 5th ed. New York: Waveland Press.

McQuoid-Mason, David. 2012. "Human Tissue and Organ Transplant Provisions: Chapter 8 of the National Act and Its Regulations, in Effect from March 2012—What Doctors Must Know." *South African Medical Journal* 102 (9): 733–35. www.samj.org.za/index.php/samj/article/view/6047.

Mencius. [1895] 2009. *The Works of Mencius.* Reprint of the edition translated by James Legge. Nothingistic Library, edited by Stephen R. McIntyre. http://nothingistic.org/library/mencius/.

Menzel, Paul T., and Bonnie Steinbock. 2013. "Advance Directives, Dementia, and Physician-Assisted Death." *Journal of Law, Medicine and Ethics* 41 (2): 484–500.

Montaigne, Michel de. 1952. "That to Philosophize Is to Learn to Die." In *The Essence of Michel Eyquem de Montaigne,* translated by Charles Cotton, 28–36. Chicago: Encyclopaedia Britannica.

Moody, Raymond. 2001. *Life after Life: The Investigation of a Phenomenon—Survival of Bodily Death.* New York: HarperOne.

Moore v. Regents of the University of California. 1990. 51 Cal. 3d 120; 271 Cal. Rptr. 146; 793 P.2d 479.

Moravec, Hans. 1990. *Mind Children: The Future of Robots and Human Intelligence.* Cambridge, MA: Harvard University Press.

Mountford, Charles P. 1965. "The Death of Jinini—An Australian Tale." In *The Dreamtime: Australian Aboriginal Myths in Paintings by Ainslie Roberts,* edited by C. P. Mountford, 28. Australia: Rigby.

Nagel, Thomas. 1979. "Death." In *Mortal Questions,* 1–10. Cambridge: Cambridge University Press.

National Association of Evangelicals. 1994. "Termination of Medical Treatment." Resolution adopted at the Annual Convention of the National Association of Evangelicals .www.euthanasia.com/cvangel.html.

National Funeral Directors Association. 2016. "Statistics." www.nfda.org/news/statistics.

National POLST Paradigm Office. 2016. "What Is POLST." www.polst.org/about-the-national-polst-paradigm/what-is-polst/.

Neuhaus, Richard John. 2000. *The Eternal Pity.* Notre Dame, IN: University of Notre Dame Press.

New Revised Standard Version Bible, Anglicized. 1989. Division of Christian Education of the National Council of the Churches of Christ in the United States of America.

https://www.biblegateway.com/versions/New-Revised-Standard-Version-Anglicised
-NRSVA-Bible/#copy.

"Nirayavagga: Hell." 2013. *Dhammapada* 22. In *Access to Insight: Readings in Theravāda
Buddhism (BCBS Edition)*. Translated from the Pali by Acharya Buddharakkhita. Last
revised November 30. www.accesstoinsight.org/tipitaka/kn/dhp/dhp.22.budd.html.

The Noble Qur'an. 2011. Translated by Muhammad Taqi-ud-Din Al-Hilali and Muhammad Muhsin Khan. Houston: Dar-us-Salam Publications.

Nuland, Sherwin. 1995. *How We Die: Reflections on Life's Final Chapter.* New York: Vintage.

Olendzki, Andrew. 2013. "Skinny Gotami and the Mustard Seed" (ThigA 10.1). In *Access to Insight: Readings in Theravāda Buddhism (BCBS Edition)*. Last revised November 30. www.accesstoinsight.org/noncanon/comy/thiga-10-01-ao0.html. Video at
https://youtu.be/Tbc9ZU1kGsA.

Oregon Death with Dignity Act. 2013. 127.800-127.897. http://public.health.oregon.gov
/ProviderPartnerResources/EvaluationResearch/DeathwithDignityAct/Pages/pas
forms.aspx.

"Pachacuti." 2017. *Wikipedia,* n.d., accessed July 17. http://en.wikipedia.org/wiki/Pacha
cuti.

Pallis, Christopher A. 2017. "Death." In *Encyclopaedia Britannica.* www.britannica.com
/EBchecked/topic/154412/death.

Parry, Bronwyn. 2004. "Technologies of Immortality: Brain on Ice." *Studies in History and
Philosophy of Biological and Biomedical Sciences* 35 (2): 391–94. http://dx.doi.org/10
.1016/j.bbr.2011.03.031.

"Patrick Stokes (Philosopher)." 2017. *Wikipedia,* n.d., accessed April 8. http://en.wikipedia
.org/wiki/Patrick_Stokes_(philosopher).

Pausch, Randy. 2008. *The Last Lecture.* www.thelastlecture.com/.

Pellegrino, Edmund D. 2000. "Decisions to Withdraw Life-Sustaining Treatment: A Moral
Algorithm." *Journal of the American Medical Association* 283 (8): 1065–67.

———. 2005. "Futility in Medical Decisions: The Word and the Concept." *HEC Forum* 17
(4): 308–18.

"The Perverted Message—A Hottentot Tale." 1963. In *African Myths and Tales*, edited by
Susan Feldmann, 107. New York: Dell.

Plato. 1953a. *Phaedo.* In *The Dialogues of Plato*, vol. 1, 4th ed., translated by Benjamin
Jowett, 414, 417–19, 474, 475–76. Oxford: Oxford University Press.

———. 1953b. *Republic.* In *The Dialogues of Plato*, vol. 2, 4th ed., translated by Benjamin
Jowett, 491–93. Oxford: Oxford University Press.

Pojman, Louis P. 2002. *Life and Death: Grappling with the Moral Dilemmas of Our Times.*
Belmont, CA: Wadsworth.

Potter, Van Rensselaer. 1970. "Bioethics: The Science of Survival." *Perspectives in Biology
and Medicine* 14:127–53.

———. 1971. *Bioethics: Bridge to the Future.* Englewood Cliffs, NJ: Prentice-Hall.

President's Commission for the Study of Ethical Problems in Medicine and Biomedical and Behavioral Research. 1981. *Defining Death: Medical, Legal and Ethical Issues in*

the Determination of Death. Washington, D.C.: US Government Printing Office. http://bioethics.georgetown.edu/pcbe/reports/past_commissions/defining_death.pdf.

President's Council on Bioethics. 2009. "A Summary of the Council's Debate on the Neurological Standard for Determining Death." In *Controversies in the Determination of Death: A White House Paper by the President's Council on Bioethics,* January, chap. 7. https://bioethicsarchive.georgetown.edu/pcbe/reports/death/chapter7.html.

Pyszczynski, Tom, Sheldon Solomon, and Jeff Greenberg. 2003. *In the Wake of 9/11: The Psychology of Terror.* Washington, D.C.: American Psychological Association.

Quindlen, Anna. 2000. *A Short Guide to a Happy Life.* New York: Random House.

Rachels, James. 1966. *The Ends of Life: Euthanasia and Morality.* Oxford: Oxford University Press.

Rahula, Walpola. 1974. *What the Buddha Taught.* New York: Grove Press.

Rando, Theresa A. 1984. *Grief, Dying, and Death: Clinical Interventions for Caregivers.* Champaign, IL: Research Press.

"Reincarnation." 2011. In *Encyclopedia of Death and Dying.* www.deathreference.com/Py-Se/Reincarnation.html.

"Reincarnation." 2017. *Wikipedia,* n.d., accessed July 23. http://en.wikipedia.org/wiki/reincarnation.

Rich, Ben A. 1997. "Postmodern Personhood: A Matter of Consciousness." *Bioethics* 11 (3–4): 206–16.

———. 2014. "Structuring Conversations on the Fact and Fiction of Brain Death." *American Journal of Bioethics* 14 (8): 31–33.

Robert Wood Foundation. 2014. *"Five Wishes": An Easy Advance Directive Promotes Dialogue on End-of-Life Care.* http://pweb1.rwjf.org/reports/grr/029110s.htm.

Rockloff, Jonathan D. 2014. "Palliative Care Gains Favor as It Lowers Costs." *Wall Street Journal,* February 23. www.wsj.com/articles/SB10001424052702303942404579363050 21497272.

Rohde, Erwin. [1921] 1925. *Psyche: The Cult of Souls and Belief in Immortality among the Greeks.* New York: Harper and Row.

Rosenberg, Alexander. 1998. *Thinking Clearly about Death.* 2nd ed. Indianapolis: Hackett.

Rudnick, Abraham. 2002. "Depression and the Competence to Refuse Psychiatric Treatment." *Journal of Medical Ethics* 28:151–55. http://jme.bmj.com/content/28/3/151.full.

"Salla Sutta: The Arrow" (*Sutta Nipata* 3.8). [1983] 1994. In *Access to Insight (Legacy Edition),* translated from the Pali by John D. Ireland. Kandy, Sri Lanka: Buddhist Publication Society.

Sarmiento de Gamboa, Pedro. 1907. "Death of Pachacuti Inca Yupanqui." In *History of the Incas,* translated by Clements Markham, 138–39. Cambridge: Hakluyt Society. www.sacred-texts.com/nam/inca/inca02.htm.

Sartre, Jean-Paul. [1943] 1956. "My Death." In *Being and Nothingness: A Phenomenological Essay on Ontology,* translated by Hazel E. Barnes, 507–29. New York: Philosophical Library.

Schneider, John M. 2001. *Grief / Depression Assessment Inventory*. www.integraonline.org /assessments/grief_depression_inventory.pdf.

Schneidman, Edwin S. 2004. *Autopsy of a Suicidal Mind*. New York: Oxford University Press.

Seneca, Lucius Annaeus. 1987. "Letter to Luciliuc, No. 70." In *The Stoic Philosophy of Seneca*, translated by M. Hadas, 202. Garden City, NY: Doubleday.

———. 2010. *Ad Lucilium Epistulae Morales*. Vol. 1. N.p.: Nabu Press. https://ryanfb .github.io/loebolus-data/L076.pdf.

"Seneca the Younger." 2017. *Wikipedia*, n.d., accessed June 30. http://en.wikipedia.org /wiki/Seneca_the_Younger.

Skloot, Rebecca. 2010. *The Immortal Life of Henrietta Lacks*. New York: Crown.

Singer, Peter. 1994. *Rethinking Life and Death*. New York: St. Martin's Press.

Smith, Huston. 1991. *The World's Religions*. New York: HarperSanFrancisco.

Soccio, Douglas J. 2004. *Archetypes of Wisdom: An Introduction to Philosophy*. 5th ed. California: Wadsworth.

Sofka, Carla J., Illene Noppe Cupit, and Kathleen R. Gilbert. 2012. "Thanatechnology as a Conduit for Living, Dying, and Grieving in Contemporary Society." In *Dying, Death, and Grief in an Online Universe*, edited by C. J. Sofka, Illene Noppe Cupit, and Kathleen R. Gilbert, 3–15. New York: Springer.

Sommers, Deborah, ed. 1995. *Chinese Religion: An Anthology of Sources*. New York: Oxford University Press.

Sprott, Richard L. 2008. "Reality Check: What Is Genetic Research on Aging Likely to Produce, and What Are the Ethical and Clinical Implications of Those Advances?" In *Aging, Biotechnology, and the Future*, edited by Catherine Y. Read, Robert C. Green, and Michael A. Smyer, 3–9. Baltimore: Johns Hopkins University Press.

Steiner, Susan. 2012. "5 Regrets of the Dying." *Guardian*, February 1. www.theguardian .com/lifeandstyle/2012/feb/01/top-five-regrets-of-the-dying.

Stevenson, Ian. 1997. *Where Reincarnation and Biology Intersect*. New York: Praeger.

Stokes, Patrick. 2012. "Ghosts in the Machine: Do the Dead Live On in Facebook?" *Philosophy and Technology* 25:363–79. www.springerlink.com/content/v258545u7v4h6407 /fulltext.pdf.

Termination of Life on Request and Assisted Suicide (Review Procedures) Act [Netherlands]. 2002. https://www.eutanasia.ws/leyes/leyholandesa2002.pdf.

"Thomas Aquinas." 2017. *Wikipedia*, n.d., accessed July 24. http://en.wikipedia.org/wiki /Thomas_Aquinas.

The Tibetan Book of the Dead. 1992. Translated with commentary by Francesca Fremantle and Chögyam Trungpa. Boston: Shambhala Press.

Tipler, Frank J. 1994. *The Physics of Immortality: Modern Cosmology, God and the Resurrection of the Dead*. New York: Doubleday.

"Topics: Physician-Assisted Suicide." 2013. On *Ethics in Clerkships* website, University of Illinois at Chicago College of Medicine. www.uic.edu/depts/mcam/ethics/suicide .htm#topic2. No longer available online.

Tsunetomo, Yamamoto. 1979. *Hagakure: The Book of the Samurai.* Translated by William Scott Wilson. Tokyo: Kodansha International (Shambhala).

Uniform Definition of Death Act. 1978. 42 U.S.C. §1802.

Valea, Ernest. 2012. "Reincarnation: Its Meaning and Consequences." *Many Paths to One Goal? A Comparative Analysis of the Major World Religions from a Christian Perspective,* June 16. www.comparativereligion.com/reincarnation.html.

Veatch, Robert M. 1976. *Death, Dying, and the Biological Revolution.* New Haven, CT: Yale University Press.

———. 1993. "The Impending Collapse of the Whole-Brain Definition of Death." *Hastings Center Report* 23 (4): 18–24.

Wahlster, Sarah, Eelco F. M. Wijdicks, Pratik V. Patel, David M. Greer, J. Claude Hemphill, III, Marco Carone, and Farrah J. Mateen. 2015. "Brain Death Declaration." *Neurology* 84:1870–79.

"Walk On—A Contemporary Cherokee Poem." 2003. In *Graceful Passages: A Companion for Living and Dying,* edited by Michael Stillwater and Gary Malkin, 32–33. Novato, CA: New World Library.

Ware, Bronnic. 2012. *The Top Five Regrets of the Dying: A Life Transformed by the Dearly Departed.* New York: Hay House.

Washington Death with Dignity Act. 2013. "Initiative 1000." www.doh.wa.gov/YouandYourFamily/IllnessandDisease/DeathwithDignityAct.aspx.

"What Are My Risk Factors?" 2015. *Time,* July 6, 70–71.

Wijdicks, Eelco F. M. 2002. "Brain Death Worldwide: Accepted Fact but No Global Consensus in Diagnostic Criteria." *Neurology* 58 (1): 20–25.

Wilkinson, Martin, and Stephen Wilkinson. 2016. "The Donation of Human Organs." In *Stanford Encyclopedia of Philosophy.* http://plato.stanford.edu/entries/organ-donation/.

Willis, Claire. 2014. *Lasting Words: A Guide to Finding Meaning toward the Close of Life.* Brattleboro, VT: Green Writers Press.

World Health Organization. 2018. "The Top 10 Causes of Death." Fact Sheet, May. www.who.int/news-room/fact-sheets/detail/the-top-10-causes-of-death.

Yalom, Irvin D. 2009. *Staring at the Sun: Overcoming the Terror of Death.* New York: Jossey-Bass.

Young, Robert. 2016. "Voluntary Euthanasia." In *Stanford Encyclopedia of Philosophy.* http://plato.stanford.edu/entries/euthanasia-voluntary/.

Youngner, Stuart, Robert M. Arnold, and Renie Shapiro, eds. 1999. *The Definition of Death.* Baltimore: Johns Hopkins University Press.

Zeldin, Wendy. 2011. "China: Case of Assisted Suicide Stirs Euthanasia Debate." *Global Legal Monitor,* August 17. Law Library of Congress. www.loc.gov/law/foreign-news/article/china-case-of-assisted-suicide-stirs-euthanasia-debate/

"Zuni." 2017. *Wikipedia,* n.d., accessed July 2. http://en.wikipedia.org/wiki/Zuni.

MARY ANN G. CUTTER

is professor of philosophy
at the University of Colorado, Colorado Springs.
She is the author and co-author of a number of books,
including *Thinking through Breast Cancer:*
A Philosophical Exploration of Diagnosis, Treatment, and Survival.

CPSIA information can be obtained
at www.ICGtesting.com
Printed in the USA
LVHW022240300719
625880LV00007B/240/P